T0074539

Digitalized and Harmonized Industrial Production Systems

Digitalized and Harmonized Industrial Production Systems

The PERFoRM Approach

Edited by

ARMANDO W. COLOMBO, MICHAEL GEPP, JOSÉ BARATA,
PAULO LEITÃO, JOSÉ BARBOSA, AND JEFFREY WERMANN

CRC Press
Taylor & Francis Group
Boca Raton London New York

CRC Press is an imprint of the
Taylor & Francis Group, an **informa** business

CRC Press
Taylor & Francis Group
6000 Broken Sound Parkway NW, Suite 300
Boca Raton, FL 33487-2742

© 2020 by Taylor & Francis Group, LLC
CRC Press is an imprint of Taylor & Francis Group, an Informa business

No claim to original U.S. Government works

Printed on acid-free paper

International Standard Book Number-13: 978-0-367-20661-1 (Hardback)

Library of Congress Cataloging-in-Publication Data

Names: Colombo, Armando W., editor.
Title: Digitalized and harmonized industrial production systems : the PERoFORM approach / edited by Armando W. Colombo, Michael Gepp, Jose Barata, Paulo Leitao, Jose Barbosa and Jeffrey Wermann.
Description: First edition. | Boca Raton, FL : CRC Press/Taylor & Francis, 2019. | Includes bibliographical references and index.
Identifiers: LCCN 2019029262 (print) | LCCN 2019029263 (ebook) | ISBN 9780367206611 (hardback ; acid-free paper) | ISBN 9780429263316 (ebook)
Subjects: LCSH: Flexible manufacturing systems. | Manufacturing processes--Automation.
Classification: LCC TS155.65 .D54 2019 (print) | LCC TS155.65 (ebook) | DDC 670.42/7--dc23
LC record available at https://lccn.loc.gov/2019029262
LC ebook record available at https://lccn.loc.gov/2019029263

Visit the Taylor & Francis Web site at
http://www.taylorandfrancis.com

and the CRC Press Web site at
http://www.crcpress.com

Contents

Abbreviations

Abbreviation	Explanation
AAS	Asset Administration Shell
API	Application Programming Interface
ARUM	Adaptive Production Management
CAD	Computer-Aided Design
CNC	Computerized Numerical Control
CoAP	Constrained Application Protocol
CPPS	Cyber-Physical Production Systems
CPS	Cyber-Physical System
DCS	Direct Control System
DDS	Data Distribution Service
DDSI-RTPS	DDS Interoperability Wire Protocol
DPWS	Devices Profile for Web Services
EMC²-Factory	Eco Manufactured transportation means from Clean and Competitive Factory
ERP	Enterprise-Resource-Planning
ESB	Enterprise Service Bus
ETCI	European Telecommunication Standards Institute
FLEXA	advanced FLEXible Automation cell
GRACE	inteGration of pRocess and quAlity Control using multi-agEnt technology
GSM	Global System for Mobile Communications
GUI	Graphical User Interface
HMI	Human Machine Interface
HTTP	Hypertext Transfer Protocol
I4.0	Industrie 4.0
ICCT	Information-Communication-Control-Technologies
ICPS	Industrial Cyber-Physical System
ICT	Information and Communication Technology
IDEAS	Instantly Deployable Evolvable Assembly Systems
IEC	International Electrotechnical Commission
IEEE	Institute of Electrical and Electronics Engineers
iESB	Intelligent Enterprise Service Bus
IIC	Industrial Internet Consortium
IIoT	Industrial Internet of Things
IMC-AESOP	ArchitecturE for Service-Oriented Process-Monitoring and - Control
IOT	Internet of Things
IP	Internet Protocol
ISA	Industry Standard Architecture

ISO	International Organization for Standardization
ITU	International Telecommunication Union
IVI	Industrial Value Chain Initiative
JDBC	Java Database Connectivity
JSON	JavaScript Object Notation
KOSF	Korea Smart Factory Foundation
KPI	Key Performance Indicator
MANUCLOUD	Distributed Cloud product specification and supply chain manufacturing execution infrastructure
MAS	Multi-Agent System
MES	Manufacturing Execution System
MIIT	Ministry of Industry and Information technology of China
MQTT	Message Queuing Telemetry Transport
NIST	National Institute of Standards and Technology
OCF	Open Connectivity Foundation
ODBC	Open Database Connectivity
OEM	Original Equipment Manufacturer
OEE	Overall Equipment Effectiveness
OPC-UA	Open Platform Communications Unified Architecture
PERFoRM	Production harmonizEd Reconfiguration of Flexible Robots and Machinery
PLC	Programmable Logic Controller
PRIME	Plug and produce intelligent multi-agent environment based on standard technology
R&D	Research and Development
RDBMS	Relational Database Management System
ReBorn	Innovative Reuse of modular knowledge Based devices and technologies for Old, Renewed and New factories
REST	Representational State Transfer
SCADA	Supervisory Control and Data Acquisition
SME	Small and Medium-sized Enterprises
SOA	Service-Oriented Architecture
SOAP	Simple Object Access Protocol
SOCRADES	Service-oriented cross-layer infrastructure for distributed smart embedded systems
SQL	Structured Query Language
TCP	Transmission Control Protocol
TSN	Time Sensitive Networks
UML	Unified Modeling Language
VDE	Verband der Elektrotechnik Elektronik Informationstechnik
VM	Virtual Machines
WS	Web Service
WSAN	Wireless Sensor and Actor Network
WSDL	Web Services Description Language
XML	Extensible Markup Language

Foreword 1

Today Manufacturing and other industrial systems have been greatly impacted by the growth of devices in Internet of Things, Artificial Intelligence, and Cyber-Physical Systems. These technologies move many manual or ad hoc processes in industrial systems to systems where failures and controls can be predicted and often executed automatically in the system. Much of the scope of this is address in the scope of Industry 4.0 industrial revolution. Cyber-Physical Systems also enables machine learning based virtual simulation of an industrial system paired to the actual system allowed for improved machine learning and prediction of events. Together these allow for significant operational improvements, and shorter time to market and overall greatly improved performance of industrial systems.

PERFoRM represents an architecture and solution strategies for a Cyber-Physical System. The authors describe the architecture, design, and provide examples of how the architecture can be applied in existing systems including test beds, compressors, electric vehicles, aerospace, and other uses. The background and samples are very useful for companies interested in implementing a Cyber-Physical System for their environment.

Michael W. Condry, PhD, FIEEE
Chief Technical Officer, Ecosystem Division, Intel Corporation (Retired), USA
President, IEEE Technology and Engineering Management Society (TEMS)
Senior Adcom, IEEE Industrial Electronics Society

Michael is currently the chair, of the Advisory Board for Clinicai, Inc. Clinicai is a BioMed startup company focused on early detections of medical symptoms such as cancer using accurate and noninvasive methodology. Michael is also currently the president of the IEEE TEMS Society. His industry experience includes startups such as Clinicai as well has leadership roles in major corporations such as Intel, Sun, and AT&T Bell Laboratories. Most recently at Intel, Michael was the chief technical officer in the Client Division. His career has a mixture of academic and industry positions, mostly in industry. Holding teaching and research positions at Princeton and University of Illinois, Urbana-Champaign. Michael came to Intel from Sun to head up Networking Applications research in Intel Labs. Michael's CTO role drove on customer innovation, design cost reduction, and other technologies and leading technical staff development. He led efforts to help the customer with system design and security. This plus efforts in technical staff development at Intel awarded him and his team the prestigious Intel Quality Award in 2015. At Sun he led the development of Unix standards as well as improved the architecture processes by engineering. At AT&T he was

one of the architects for the BellMac 32 processor, the first 32 bit micropro-
cessor on the market and lead software projects including a Real-Time Unix
design and Unix System V file system. His background includes projects in
computer architecture, software, firmware, operating systems, networking,
IoT, internet applications, standards, and computer security. Michael retired
from Intel in June 2015.

Michael has patents in computer architecture and security. He has published many technical papers and regularly presents keynotes at technical
conferences.

Michael has many years engaging in the IEEE. He is the President of the
IEEE Technology and Engineering Management Society (TEMS). Michael
is a senior board member for the IEEE Industrial Electronics Society (IES);
he created and chairs the IEEE Industry Forum series that has successfully
engaged industry in over 18 conferences. Michael is also a member of the
IEEE Computer Society for over 28 years. He has chaired many IEEE conferences as well as the Industry Forum program in multiple societies.

Foreword 2

Cyber-Physical Systems (CPSs) study the deep integration between cyber information systems and dynamic physical systems. The domains of those physical systems include energy systems, automotive systems, aerospace systems, etc., many of which are addressed in different chapters of this book.

The key to design a high quality CPS is to identify the deep underlying links between the information system and the target physical system. In short, one needs to develop a cyber-physical solution to tackle a cyber-physical problem. In a typical CPS context, an incomplete physical model and a set of data are given. The development of a cyber-physical solution is essentially to make the incomplete physical model "more complete." This requires the confluence of the physical model based approach and the data driven approach. Such a confluence identifies the underlying links between the information system and the physical system, and characterizes the cyber-physical solution. It pushes the envelope in the modeling of physical problems and the development of tools, through developing large-scale analytical modeling techniques, high performance simulation, synthesis and verification techniques, advanced data analytics techniques, etc. In addition to data, tools, and models, CPS design needs to address other important factors such as security, architecture, timing, power, and reliability. Considering the above aspects, while exploring the unique nature of the target physical domain, clearly requires the close collaborations among information technology experts and physical domain experts. This book provides a set of excellent examples demonstrating such collaborations. It is my firm belief that the research dedicated to the highly interdisciplinary CPS research will keep growing and eventually take it to a new level.

Shiyan Hu
Michigan Technological University
Houghton, MI, USA

Professor Shiyan Hu received his PhD in Computer Engineering from Texas A&M University in 2008. He is an associate professor at Michigan Technological University, and he was a visiting associate professor at Stanford University. His research interests include Cyber-Physical Systems (CPS), CPS Security, and Data Analytics where he has published more than 100 refereed papers.

Prof. Hu is an ACM Distinguished Speaker, an IEEE Systems Council Distinguished Lecturer, a recipient of U.S. National Science Foundation CAREER Award, and a recipient of IEEE Computer Society TCSC Middle Career Researcher Award. His research was highlighted as a Keynote Paper in IEEE Transactions on Computer-Aided Design in 2017, as a Thomson Reuters ESI Highly Cited Paper in 2017, and as the Front Cover in IEEE Transactions on Nanobioscience in March 2014.

Prof. Hu is the Chair for IEEE Technical Committee on Cyber-Physical Systems. He is the Editor-In-Chief of IET Cyber-Physical Systems: Theory & Applications. He has been an Associate Editor for IEEE Transactions on Computer-Aided Design, IEEE Transactions on Industrial Informatics, IEEE Transactions on Circuits and Systems, ACM Transactions on Design Automation for Electronic Systems, and ACM Transactions on Cyber-Physical Systems. He is also a Guest Editor for eight IEEE/ACM journals such as Proceedings of the IEEE (PIEEE) and IEEE Transactions on Computers. He has held chair positions in various IEEE/ACM conferences. He is a Fellow of IET.

Foreword 3

Enabling the Industrial Digital Transformation

In fast-paced business environments, the modern enterprises need to stay competitive. To do so, they continuously aim for optimization of their business processes as well as interactions with other stakeholders. Digitalization, as envisioned by the fourth industrial revolution, poses another challenge that needs to be effective and in a timely fashion mastered.

However, this is easier said than done. The disruptive changes that need to be implemented, in conjunction with the introduction of new technologies, modi operandi, business models, and stakeholder interaction make the whole undertaking a challenging issue. A lot of marketing hype often overestimates the expected benefits and the timeline that they can be achieved. Furthermore, the change management process and its complexities toward adjusting to the new era are for many large companies, and even more for small- and medium-sized enterprises not clear and tangible.

To embrace the benefits of Industrie 4.0 and carry out effectively the necessary digital transformation, the changes need to be understood, well-planned, measured, and successfully executed in a way that fully addresses the complexities underlying in their sociotechnical dimensions. For the technical side, this implies a good understanding of the existing as well as planned technology adoption, in order to design and realize architectures that fully capitalize on the Industrie 4.0 capabilities from the production systems up to enterprise systems level.

In this process, the engineering and the development of tools are expected to play a pivotal role. To have a successful digital transformation, it is indispensable to address the migration paths and strategies for systems, processes, and personnel that utilize them. New tools that enable such efforts to help the enterprise to optimize its resource usage, and adjust its processes for the new infrastructure need to be developed. Coordinating all of these actions in an efficient manner, ultimately empowers modern enterprises to stay competitive enhance their performance.

In this new exciting era, this book touches on several aspects of both architectural as well as engineering nature that need to be tackled in the manufacturing domain. By demonstrating how the PERFoRM framework addresses several of the posed requirements and associated challenges, a clearer path emerges on what actions need to be undertaken and how to carry them out. In addition, the book provides several industrial application examples, for example, from compressors, white goods, electric vehicles, and aerospace, which exemplify how the PERFoRM framework was utilized and make even more clear the hurdles and benefits that can be seized.

Stamatis Karnouskos
SAP
Industrie 4.0 & Digital Supply Chain Innovation
Dietmar Hopp Allee 16
Walldorf, Germany

Stamatis Karnouskos is an expert on the Internet of Things at SAP, Germany. He investigates the added value and impact of emerging technologies in enterprise systems. For over 20 years, he has led efforts in several European Commission and industry-funded projects related to the Internet of Things, Cyber-Physical Systems, Industrie 4.0, manufacturing, smart grids, smart cities, security, and mobility. Stamatis has extensive experience on research and technology management within the industry as well as the European Commission and several national research funding bodies (e.g., in Germany, France, Switzerland, Denmark, Czech Republic, and Greece). He has served on the technical advisory board of Internet Protocol for Smart Objects Alliance (IPSO), and the Permanent Stakeholder Group of the European Network and Information Security Agency (ENISA).

Preface

Industrial competitiveness today means shorter product lifecycles, increased product variety, shorter time-to-market, shorter time-in-market, and customized tangible products and services. To face these challenges, the manufacturing industry is supported by novel paradigms and associated technologies, namely Industry 4.0, Industrial Internet of Things, and Industrial Cyber-Physical Systems. The penetration of these groundbreaking principles in the industrial manufacturing ecosystem push them to move from traditional hierarchical management, control and automation approaches toward digitalized and informatized, reconfigurable, and networked flexible manufacturing systems that are both, structurally reconfigurable and evolvable, and functional (i) dynamically adaptable to react to changing production environment and (ii) flexible to different business opportunities.

For years several emergent engineering approaches and related Information-Communication-Control-Technologies (ICCT), such as Multi-Agent Systems, Service-oriented Architecture, Plug-and-Produce systems, Cloud and Fog technologies, Smart Big Data and Analytics, among others, have been researched and prototypes developed in a variety of research and innovation activities. The confluence of those results with the latest developments in the industrial digitalization sector, for example, Digitalized Mechatronics, Industrial Cyber-Physical Systems, Systems-of-Cyber-Physical Systems (SoCPS) Engineering, Internet of Things, Industry 4.0, and Internet of Services, is opening a new broad spectrum of innovation possibilities for researchers, practitioners, and industrialists. The PERFoRM (Production harmonizEd Reconfiguration of Flexible Robots and Machinery) approach deals with this confluence and brings the mentioned developments to a new level of technology readiness with selected industrial application.

This book introduces the vision and describes the major results of research, development, and innovation works carried out by several major industrial players, leading universities, and research institutions, within the European HORIZON2020 Factory-of-the-Future project "Production harmonizEd Reconfiguration of Flexible Robots and Machinery." More specifically, PERFoRM consists of a Research and Innovation approach that covers several aspects: (1) understanding the major characteristics of industrial manufacturing systems when they are digitalized following the Industry 4.0 paradigm with a special focus on the improvement of reconfigurability and flexibility of production systems; (2) showing, by prototype implementing in real industrial environments, the novel results of digitalizing production machinery and robots, by applying Industrial Cyber-Physical Systems related technologies; (3) understanding what happens when machinery and robots are networked using Industrial Internet-of-Things Technology and can talk about products and production with each other and with the human using the Internet of Services Technology; and (4) evaluating the impact of those innovations in a broad scientific and industrial community.

This book (1) reviews the state-of-the-art methodologies and technologies of flexible, reconfigurable, digitalized, and informatized manufacturing systems, outlines the innovations beyond that state that are necessary to achieve PERFoRM and describes arising benefits of implementing the PERFoRM approach; (2) presents new innovations addressing architectural and functional aspects of digitalized industrial manufacturing systems, as well as new engineering methods and tools to support the migration of legacy systems into Industry 4.0-compliant manufacturing infrastructures; (3) describes a set of test beds and industrial use cases in four different application domains (electric vehicles, white goods, aerospace, and machine production) that are used to validate the PERFoRM approach; and (4) overviews current trends and challenges in industrial digitalization in front of the proposed PERFoRM approach, summarizing the impact of the different innovated solutions, and out looking new research and innovation opportunities beyond those latest innovations.

Going through the following pages, the reader will get a deeper view of the PERFoRM approach from multiple angles:

- Chapter 1 is dedicated to the introduction to what the reader can expect being presented within this book. It provides an overview: (i) of the motivation; (ii) of the state-of-the art; (iii) of the industrial requirements; (iv) of the efforts carried out by the partners of the PERFoRM project, toward defining the vision of Digitalized and Harmonized Industrial Production Systems and demonstrating its advantages in front of today's situation and trends for implementing reconfigurable industrial manufacturing systems within the context of Industry 4.0-compliant engineering of manufacturing systems.

- Chapter 2 focuses on assessment of promising technologies and standards available in an industrial context including a discussion about challenges for research and innovation activities. While investigating the introduction of digitalization in industrial manufacturing systems, and even from the shop floor (OT) to upper levels of an enterprise (IT), latest standards, proved technologies, industrial solutions, and latest research and innovation works in the industrial automation and digitalization have been considered.

- Chapter 3 introduces in a stepwise manner the PERFoRM architecture, positioning it in front of standard management, control, and automation enterprise architectures. Within this chapter, a specific PERFoRM-compliant system architecture is proposed attempting to cover the basic needs for implementing the PERFoRM approach in four different industrial scenarios.

- Chapter 4 deals with the aspect "Data and Information Inter-operability" that is essential for guaranteeing the functional integration of all components of the PERFoRM architecture. It introduces the major specifications of a PERFoRM data model "PERFoRMML" defined on the basis of the standard AutomationML and developed for being later used in the test beds and industrial use cases.

- Chapter 5 focuses on the "Connectivity" between components located on the "Operational Technology (OT)" level and those situated on the "Information Technology (IT)" level of the industrial production systems. In order to facilitate the integration of legacy HW and SW production components and systems, a series of customized "Adaptors" were developed as part of the PERFoRM approach, responsible to translate the device/module internal data model into the one introduced in Chapter 4. The Chapter 5 presents the definitions, the specifications, and explains the main features of the implementation of those Technology Adaptors belonging to the PERFoRM architecture.

- Chapter 6 is dedicated presenting the definition, explaining the role within the PERFoRM approach, and describing the major specifications of a PERFoRM middleware component. It summarizes details of the prototype implementations of this central component of the PERFoRM architecture.

- Chapter 7 deals with engineering methods and tools. These are seen as key enablers for efficiently designing, testing, deploying, and operating any PERFoRM-compliant industrial manufacturing system. An appraisal of existing engineering methods and tools, appropriate to the test beds and the four industrial use cases, is presented, followed by the description of the tools developed based on the PERFoRM approach.

- Chapter 8 focuses on the stepwise introduction of the PERFoRM approach into the current industrial manufacturing ecosystem, which requires a systematic approach to migrate from legacy systems into a digitalized, harmonized, reconfigurable, and flexible production system. The migration procedure, consisting on a migration strategy complemented with a migration plan and a migration process aims to preserve the functional integration and to maintain the performance aspects of the digitalized systems, considering also the possible coexistence of the PERFoRM-compliant solutions and legacy (non-digitalized) components of the industrial system.

  To better depict the innovations achieved with the application of the PERFoRM approach, a series of five chapters are dedicated to present a set of two test beds and four industrial field trials, use cases, demonstrating and evaluating the results of implementing PERFoRM solutions that achieved levels 5, 6, and 7 of technology readiness respectively.

- Chapter 9 discusses how prior to the actual deployment of the PERFoRM-compliant solutions at the premises of the four industrial end users (industrial use cases), the basic principles of the PERFoRM system architecture and technologies were validated within the industry-like environments of four pretest beds implemented at Loccioni, Italy, Technical University Braunschweig, Germany, Polytechnic of Bragança, Portugal, and Institute I2AR in Emden, Germany, and two test beds, one set at the Manufacturing Technology Centre (MTC) in Coventry, UK, and another one at the Technologie-Initiative SmartFactory KL e.V. located in Kaiserslautern, Germany. Here the reader will explore how the work performed on the test beds did allow the verification and validation of the technological developments and the confirmation that user requirements are met, including an overview about how the solutions are de-risked before they are implemented in industrial operational environments.

- In Chapter 10, we learn how the availability and reliability of production machinery installed at the Duisburg Compressor Factory of Siemens AG, Germany, can be improved by the integration of the PERFoRM middleware on the one hand, and on the other hand, by implementing PERFoRM-compliant tools. The chapter summarizes the innovation work realized to respond to the following two major industrial requirements defined for this use case: (1) raised machine availability through improved condition knowledge and maintenance activities and (2) better planning and scheduling of necessary downtimes of machinery for maintenance in order to reduce negative impacts on production figures (e.g., completion dates, downtimes, or schedule deviations).

- Chapter 11 describes how applying the PERFoRM approach to a Whirlpool factory located in Biandromo, Italy, which produces built-in appliances (microwave ovens), it is possible to realize a real-time monitoring system able to correlate dynamic behavior of the factory to its Key Performance Indicators (KPIs) implementation and to static indicators such as Key Business Factors (KBF). Moreover, the chapter describes the application of the PERFoRM approach for supporting the reconfiguration of the factory at shop floor level using cobot.

- Chapter 12 describes the research and innovation results, the methodologies, and technologies that have been developed and implemented by applying the PERFoRM approach to a Microfactory of IFEVS, Italy, dedicated to the production of different Electric Vehicle architectures, aiming at establishing the conditions for a near future smart and flexible manufacturing of Urban EVs.

- Chapter 13 discusses how the PERFoRM approach support building and operating a "Micro-Flow cell" at GKN Aerospace, Sweden. The chapter describes the stepwise implementation of a reconfigurable and flexible production cell and summarizes the results of validating the technical, functional and business specifications that are behind the major industrial requirements for the reconfigurability and the flexibility of the installation.

- Finally, Chapter 14 rounds up the book, arguing the application of the PERFoRM approach to support the development and industrial implementation, to levels 6 and 7 of technology readiness, of a new generation digitalized, agile, and flexible manufacturing system, based among others on plug-and-produce concept, industrial Cyber-Physical Systems, and Industry 4.0-compliant technologies. It summarizes new innovations addressing architectural and functional aspects of digitalized industrial manufacturing systems, as well as new engineering methods and tools to support the migration of legacy systems into Industry 4.0-compliant manufacturing infrastructures.

The transformation toward digitalized, harmonized, flexible, and reconfigurable manufacturing systems following the PERFoRM approach must actively be shaped to fully achieve the promised benefits. This means a considerable amount of technological, industrial application-oriented, and human-oriented challenges has to be tackled. This book identifies some of those challenges, detected, and indicates gaps that should be addressed by follow-up research, development, and innovation activities.

We hope you enjoy this book, which should inspire you to further advance the bold vision presented here, so that one day in the near future it may represent an industrial reality.

Armando W. Colombo
(I2AR, University of Applied Sciences Emden/Leer)

Michael Gepp
(Siemens)

José Barata
(UNINOVA)

Paulo Leitão
(Polytechnic Institute of Bragança)

José Barbosa
(Polytechnic Institute of Bragança)

Jeffrey Wermann
(I2AR, University of Applied Sciences Emden/Leer)

Acknowledgments

The authors would like to thank for their support the European Commission, and all the partners of the EU Horizon 2020 project PERFoRM (http://www.horizon2020-perform.eu/). The PERFoRM book in your hands has been possible due to the direct or indirect work of several people who contributed fruitful ideas, discussions, experiments, guidance, etc., and we would like to acknowledge them.

Editors

Prof. Dr.-Ing. Armando Walter Colombo (Fellow IEEE) joined the Department of Electrotechnical and Industrial Informatics at the University of Applied Sciences Emden/Leer, Germany, became full professor in August 2010 and director of the Institute for Industrial Informatics, Automation and Robotics (I2AR) in 2012. He worked during the last 17 years as manager for Collaborative Projects and also as Edison Level 2 Group Senior Expert at Schneider Electric, Industrial Business Unit.

Prof. Colombo received the BSc on Electronics Engineering from the National Technological University of Mendoza, Argentina, in 1990, the MSc on Control System Engineering from the National University of San Juan, Argentina, in 1994, and the Doctor degree in Engineering from the University of Erlangen-Nuremberg, Germany, in 1998. From 1999 to 2000 he was Adjunct Professor in the Group of Robotic Systems and CIM, Faculty of Technical Sciences, New University of Lisbon, Portugal.

Prof. Colombo has extensive experience in managing multicultural research teams in multiregional projects. He has participated in leading positions in many international research and innovation projects, managing an R&D&I total (European Union authorized) budget of ~30,00 Mio EURO in the last 12 years.

His research interests are in the fields of industrial Cyber-Physical Systems, industrial digitalization, system-of-systems engineering, Internet of Services and engineering of Industry 4.0-compliant solutions.

Prof. Colombo has over 30 industrial patents and more than 300 per-review publications in journals, books, chapters of books, and conference proceedings (see https://scholar.google.de/citations?user=FgFDTMEAAAAJ&hl=en). With his contributions, he has performed scientific and technical seminal contributions that are nowadays being used as one of the basis of what is recognized as "The 4th Industrial Revolution": networked collaborative smart Cyber-Physical Systems that are penetrating the daily life, producing visible societal changes and impacting all levels of the society.

He is member of the International Program Committee (IPC) of several scientific events, and served as general co-chair of several international conferences. Prof. Colombo is cofounder of the IEEE IES TC on Industrial Cyber-Physical Systems, TC on Industrial Agents and TC on Industrial Informatics. He is currently member of the IEEE IES Administrative Committee (AdCom) and of the IEEE Systems Council Administrative Committee (AdCom).

Prof. Colombo serves/d as advisor/expert for the definition of the R&D&I priorities within the Framework Programs FP6, FP7, and FP8 (HORIZON 2020) of the European Union, and he is working as expert in the European Research

Executive Agency (REA), ECSEL, Eureka-, Canada Digital Supercluster-, German BMBF/DLR IKT-Programs and several other national research funding bodies.

Prof. Colombo is listed in Who's Who in the World/Engineering 99-00/01 and in Outstanding People of the XX Century (Bibliographic Centre Cambridge, UK).

Dr. Michael Gepp, after acquiring his Diploma degree in mechanical engineering and business administration from the Friedrich-Alexander University of Erlangen-Nuremberg (Germany) in 2009, started his PhD thesis at Siemens Corporate Technology on economic analyses of engineering in the engineer-to-order business. Gepp acquired his PhD degree in 2014 and is now working as a technical consultant at Siemens Corporate Technology, Research Group on Systems Engineering—Engineering Methods and Tools.

His fields of interest are system engineering and Industrie 4.0 in the engineer-to-order business. In this context he is currently working on optimization of engineering processes, for example, by developing and applying modularization and standardization approaches for complex industrial systems in various research and consulting projects.

José Barata Oliveira is a professor at the Department of Electrical Engineering of the New University of Lisbon and a senior researcher of the UNINOVA Institute. He has a PhD degree in Robotics and Integrated Manufacturing since 2004 from the New University of Lisbon. Prof. Oliveira has participated in more than 15 International Research projects involving different programs (NMP, IST, ITEA, ESPRIT. HORIZON2020). Since 2004, he has been leading the UNINOVA participation in EU projects, namely EUPASS, Self-Learning, IDEAS, PRIME, RIVERWATCH, ROBO-PARTNER, PROSECO, and PERFoRM. His main research interests are in the area of Intelligent Manufacturing with particular focus on Complex Adaptive Systems, involving intelligent manufacturing devices. In the last years, he has participated actively researching SOA based approaches for the implementation of Intelligent Manufacturing Devices. He has published over 200 original papers in international journals and international conferences. He is a member of the IEEE technical committees on Industrial Agents and Industrial Cyber-Physical Systems (IES), Self-Organization and Cybernetics for Informatics (SMC), and Education in Engineering and Industrial Technologies (IES). He is also a member of the IFAC technical committee 4.4 (Cost Oriented Automation).

Paulo Leitão received the MSc and PhD degrees in Electrical and Computer Engineering, both from the University of Porto, Portugal, in 1997 and 2004, respectively. He joined the Polytechnic Institute of Bragança, Portugal, in 1995, where he is professor at the Department of Electrical Engineering. He served as head of the Department of Electrical Engineering from 2009 to 2015, vice president of Directive Board of School of Technology and Management from

2004 to 2009, president of the Pedagogical Council of School of Technology and Management during 2000, and vice president of Scientific Council of School of Technology and Management from 2001 to 2004.

His research interests are in the field of intelligent and reconfigurable systems, Cyber-Physical Systems, Internet of Things, distributed data analysis, factory automation, multi-agent systems, holonic systems, and self-organized systems. From 1993 to 1999, he developed research activities at the CIM Centre of Porto, from 1999 to 2000 at IDIT—Institute for Development and Innovation in Technology, from 2009 to 2017 at LIACC—Artificial Intelligence and Computer Science Laboratory, and since 2018 he has been the Head of CeDRI—Research Centre in Digitalization and Intelligent Robotics.

Dr. Leitão participates/has participated in several national and international research projects (e.g., EU H2020 GO0D MAN—aGent Oriented Zero Defect Multi-stage mANufacturing, EU H2020 PERFoRM—Production harmonizEd Reconfiguration of Flexible Robots and Machinery, EU FP7 ARUM—Adaptive Production Management, EU FP7 GRACE—InteGration of pRocess and quAlity Control using multi-agEnt technology, EU DEDEMAS—Decentralized Decision Making and Scheduling, EU MOSCOT—A Modular Shop Control Toolkit for Flexible Manufacturing, EU PASO—Paradigm Independent Shop Control for Smaller Manufacturing Sites, and EU CCP—CIM Centre of Porto), and Networks of Excellence (e.g., IMS—Intelligent Manufacturing Systems, and CONET—Cooperating Objects Network of Excellence). He is member of the International Program Committee (IPC) of several scientific events, and served as general cochair of several international conferences, namely IEEE INDIN'18, SOHOMA'16, IEEE ICARSC'16, HoloMAS'11, and IFAC IMS'10. He has published four books and more than 200 papers in high-ranked international scientific journals and conference proceedings (per-review). He is coauthor of three patents and received four paper awards at INCOM'06, BASYS'06, IEEE INDIN'10, and INFOCOMP'13 conferences.

Dr. Leitão is a senior member of the IEEE Industrial Electronics Society (IES) and Systems, Man and Cybernetics Society (SMCS), past chair of the IEEE IES Technical Committee on Industrial Agents, and member at-large of the IEEE IES Administrative Committee (AdCom). Currently, he is also chair of the IEEE Standards Association P2660.1 Working Group.

José Barbosa has a PhD in Automation and Computer Science from the University of Valenciennes and Hainaut-Cambrésis (France). He is a senior researcher at Polytechnic Institute of Bragança, Portugal, participating in several European funded projects, namely in the EU FP7 ARUM (Adaptive Production Management), in the EU FP7 GRACE (Integration of Process and Quality Control Using Multi-agent Technology) project and in the EU H2020 ERASMUS + DA.RE (Data Science Pathways to reimagine education), EU H2020 GO0DMAN (aGent Oriented Zero Defect Multi-stage mANufacturing), EU H2020 PERFoRM (Production harmonizEd Reconfiguration of Flexible Robots and Machinery). He is also an invited professor at the Department of Electrical Engineering of the Polytechnic Institute of Bragança, Portugal. Barbosa has more than 50 papers published at international journals and proceedings of international conferences. His main research topics focus on the development of self-organizing and evolvable manufacturing control architectures following the holonic and multi-agent system paradigms enriched with biological inspired mechanisms, particularly applied into Cyber-Physical Systems. He is also member of several IEEE and IFAC Technical Committees.

Jeffrey Wermann finished his Bachelor's degree in "Elektrotechnik und Automatisierungstechnik" (electronics & automation) at the University of Applied Sciences Emden/Leer, Germany, in 2010, followed by a Master's degree in Industrial Informatics. During his studies he was already heavily involved in research activities in topics related to real-time data processing and modern paradigms in automation, like ICPS, agent- and SOA-based systems. After finishing his studies in November 2012, he stayed at the university, becoming a scientific assistant and a member of the institute for Industrial Informatics, Automation and Robotics (I2AR). In this position he is responsible for the technical coordination of the research activities of the institute I2AR.

He is both actively taking part and supervising project work related to the developments performed in the university's main research facilities, the "Digital Factory" and the "Flexible Manufacturing Model." Besides his activities for the institute I2AR, he was involved in the EU FP7 CP "IMC-AESOP" from May 2013 until the projects end in December 2013. From October 2015, he joined the EU HORIZON2020 project "PERFoRM," being involved as project manager at the Institute I2AR in charge of the development of an industrial Middleware, which is being used as an ICT processing tool to support the connectivity and interoperability of the different solutions developed within that innovation project. Besides his research and development works, backed up with a considerable number of scientific and technical publications, Wermann is also performing graduated and postgraduated teaching activities as Invited Lecturer in the courses of the Master on Industrial Informatics, specialization "Industrial Cyber-Physical Systems" of the University of Applied Sciences Emden/Leer, Germany.

Contributors

Giacomo Angione
Research for Innovation
Loccioni
Angeli di Rosora, Italy

José Barata
Centre of Technology and Systems
UNINOVA – Instituto de
   Desenvolvimento de Novas
   Tecnologias
Caparica, Portugal

José Barbosa
Department of Electrical
   Engineering
Instituto Politécnico
   de Bragança
Bragança, Portugal

Marco Biasiotto
System Integration
Interactive Fully Electrical
   Vehicles SRL
Sommariva del Bosco, Italy

Waldemar Borsych
Institute for Informatics,
   Automation and
   Robotics (I²AR)
University of Applied Sciences
   Emden/Leer
Emden, Germany

Filippo Boschi
Department of Management,
   Economics and Industrial
   Engineering
Politecnico di Milano
Milan, Italy

Lennart Büth
Institute for Machine Tools and
   Production Technology
Technische Universität
   Braunschweig
Braunschweig, Germany

Ana Cachada
Department of Electrical
   Engineering
Instituto Politécnico
   de Bragança
Bragança, Portugal

Ambra Calà
Department of Corporate
   Technology (CT)
Siemens AG
Erlangen, Germany

Nandini Chakravorti
Data & Information
   Systems
The Manufacturing Technology
   Centre (MTC)
Coventry, United Kingdom

Armando Walter Colombo
Institute for Informatics,
   Automation and
   Robotics (I²AR)
University of Applied Sciences
   Emden/Leer
Emden, Germany

Cristina Cristalli
Research for Innovation
Loccioni
Angeli di Rosora, Italy

Pietro Cultrona
Advanced Engineering
Comau SPA
Grugliasco, Italy

Sandro De Pasquale
System Integration
Interactive Fully Electrical
   Vehicles SRL
Sommariva del Bosco, Italy

Evangelia Dimanidou
Controls & Connectivity
The Manufacturing Technology
   Centre (MTC)
Coventry, United Kingdom

Simon Eggimann
Paro AG
Subingen, Switzerland

Paola Fantini
Department of Management,
   Economics and Industrial
   Engineering
Politecnico di Milano
Milan, Italy

Adriano Ferreira
Department of Electrical
   Engineering
Instituto Politécnico
   de Bragança
Bragança, Portugal

Krister Floodh
Research and Technology Centre
GKN Aerospace Sweden AB
Trollhättan, Sweden

Matthias Foehr
Department of Corporate
   Technology (CT)
Siemens AG
Erlangen, Germany

Stefan Forsman
Research and Technology Centre
GKN Aerospace Sweden AB
Trollhättan, Sweden

Martin Frauenfelder
Paro AG
Subingen, Switzerland

Michael Gepp
Department of Corporate
   Technology (CT)
Siemens AG
Erlangen, Germany

Frederik Gosewehr
Institute for Informatics,
   Automation and Robotics (I²AR)
University of Applied Sciences
   Emden/Leer
Emden, Germany

Marco Grosso
System Integration
Interactive Fully Electrical Vehicles SRL
Sommariva del Bosco, Italy

André Hennecke
Technologie Initiative SmartFactory KL
Kaiserslautern, Germany

Manfred Hucke
XETICS GMBH
Stuttgart, Germany

Riccardo Introzzi
System Integration
Interactive Fully Electrical
   Vehicles SRL
Sommariva del Bosco, Italy

Massimo Ippolito
Advanced Engineering
Comau SPA
Grugliasco, Italy

Paulo Leitão
Department of Electrical
  Engineering
Instituto Politécnico de Bragança
Bragança, Portugal

Giulia Lo Duca
Research for Innovation
Loccioni
Angeli di Rosora, Italy

Niels Lohse
Wolfson School of Mechanical
  and Manufacturing
  Engineering
Loughborough University
Loughborough, United Kingdom

Olga Meyer
Competence Center Digital Tools for
  Manufacturing
Fraunhofer Institute for
  Manufacturing Engineering and
  Automation (IPA)
Stuttgart, Germany

Benjamin Neef
Institute for Machine Tools and
  Production Technology
Technische Universität
  Braunschweig
Braunschweig, Germany

Birgit Obst
Department of Corporate
  Technology (CT)
Siemens AG
Munich, Germany

Phil Ogun
Wolfson School of Mechanical
  and Manufacturing
  Engineering
Loughborough University
Loughborough, United Kingdom

Arnaldo Pagani
Whirlpool Operation Center
  of Excellence
Whirlpool Europe
Comerio, Italy

Davide Penserini
System Integration
Interactive Fully Electrical Vehicles SRL
Sommariva del Bosco, Italy

Ricardo Silva Peres
Centre of Technology and Systems
UNINOVA – Instituto de
  Desenvolvimento de Novas
  Tecnologias
Caparica, Portugal

Pietro Perlo
System Integration
Interactive Fully Electrical Vehicles SRL
Sommariva del Bosco, Italy

Pierluigi Petrali
Whirlpool Operation Center
  of Excellence
Whirlpool Europe
Comerio, Italy

Flávia Pires
Department of Electrical Engineering
Instituto Politécnico de Bragança
Bragança, Portugal

Sergio Pozzato
System Integration
Interactive Fully Electrical
  Vehicles SRL
Sommariva del Bosco, Italy

Steffen Raasch
Institute for Informatics,
Automation and Robotics (I²AR)
University of Applied Sciences
  Emden/Leer
Emden, Germany

Mostafizur Rahman
Data & Information Systems
The Manufacturing Technology
    Centre (MTC)
Coventry, United Kingdom

Greg Rauhöft
Competence Center Digital Tools
    for Manufacturing
Fraunhofer Institute for
    Manufacturing Engineering and
    Automation (IPA)
Stuttgart, Germany

André Dionisio Rocha
Centre of Technology and Systems
UNINOVA – Instituto de
    Desenvolvimento de Novas
    Tecnologias
Caparica, Portugal

Nelson Rodrigues
Department of Electrical Engineering
Instituto Politécnico de Bragança
Bragança, Portugal

Gioele Sabato
System Integration
Interactive Fully Electrical
    Vehicles SRL
Sommariva del Bosco, Italy

Daniel Schel
Competence Center Digital Tools
    for Manufacturing
Fraunhofer Institute for
    Manufacturing Engineering
    and Automation (IPA)
Stuttgart, Germany

Daniel Stock
Competence Center Digital Tools
    for Manufacturing
Fraunhofer Institute for
    Manufacturing Engineering and
    Automation (IPA)
Stuttgart, Germany

Marco Taisch
Department of Management,
    Economics and Industrial
    Engineering
Politecnico di Milano
Milan, Italy

Giacomo Tavola
Department of Management,
    Economics and Industrial
    Engineering
Politecnico di Milano
Milan, Italy

Sebastian Thiede
Institute for Machine Tools and
    Production Technology
Technische Universität
    Braunschweig
Braunschweig, Germany

Nils Weinert
Department of Corporate
    Technology (CT)
Siemens AG
Munich, Germany

Jeffrey Wermann
Institute for Informatics,
    Automation and
    Robotics (I²AR)
University of Applied Sciences
    Emden/Leer
Emden, Germany

Per Woxenius
Research and Technology
    Centre
GKN Aerospace Sweden AB
Trollhättan, Sweden

Disclaimer

The information and views set out in this publication are solely those of the author(s) and do not necessarily reflect the official opinion of their associated affiliation. Neither the companies, nor institutions, and bodies nor any person acting on their behalf may be held responsible for the use which may be made of the information contained therein. We explicitly note that this document may contain errors, inaccuracies, or errors or omissions with respect to the referenced materials.

1

PERFoRM: Industrial Context and Project Vision

Filippo Boschi, Giacomo Tavola, Marco Taisch
(Politecnico di Milano)

Michael Gepp, Matthias Foehr
(Siemens AG)

Armando W. Colombo
(Institute for Informatics, Automation and Robotics (I²AR))

CONTENTS

1.1 Introduction

1.1.1 Industrial Trends

The whole world economy, and naturally also the European one, is facing enormous pressures, especially in the manufacturing sector as the current production environment has to cope with many changes and challenges. One of the most relevant issues is represented by the market changes that become more frequent and unpredictable than before (Bi et al. 2008; Malhotra, Raj, and Arora 2010). In fact, modern markets are characterized by shorter product life cycles, increased product variety and shorter time to market, and by the increasing demand of responsiveness to changes in product and production volumes. Allied to these disturbances, there is also the constant need for cost production and energy consumption reduction that are going to have an important influence on new manufacturing production system trends.

Consequently, to cope with these requirements, all industrial end user scenarios need to adapt to shorter product life cycles and to reconfigure more frequently their production systems to offer new products variants, while maintaining high-quality standards, minimizing costs, and taking into account resource and energy efficiency, urban production, and demographic change.

At the same time, manufacturing enterprises are facing substantial challenges with regard to disruptive concepts enabled by several digital technologies such as the Internet of Things (IoT), and cloud-based manufacturing, service-oriented architectures (SOAs), Web services technologies, plug-and-produce systems, and cyber-physical systems (CPS) also referred to as Industry 4.0. In fact, it is not a news that the today's manufacturing domain is standing on the cusp of the fourth industrial revolution, the one which promises to marry the worlds of production and network connectivity, opening to a new age of technological innovation and outstanding performances. Among many, the CPS technologies have been considered the new manufacturing paradigm's enablers as they are intelligent systems that smartly intertwine the physical factory floor with the cyber computational space, to sense the changing state of the real world and enable a flawless enterprise control and optimization. The implementation of such technologies, Industry 4.0, which is about digital innovation in products, processes, and business models, offers opportunities that industries cannot afford to miss (Colombo et al. 2017).

In other words, the evolution of technology is pushing and it is meeting the current manufacturing trends. In particular, digital technologies are the core driver for the manufacturing transformation in the age of "Industry 4.0." In fact, the introduction of such technologies promises companies to find solutions capable of turning increasing complexity into opportunities for ensuring sustainable competitiveness and profitable growth. These technologies are able to smartly create networks related to strategic and operating values in the intelligent manufacturing environment, forcing companies to totally rethink the way they do business. They include horizontal integration of data flow between partners, suppliers, and customers, as well as vertical integration within the organization frames—from design and specification through development to final product and maintenance including recycling phases of the product and system life cycles (ARC Advisory Group 2018; Hozdić 2015; McKinsey & Company 2015). They facilitate the possibility to satisfy the continuous evolvement of market requirement.

For these reasons, the traditional approach to manufacturing control systems based on centralized or hierarchical control structures presents good characteristics in terms of productivity and optimization, but it needs to be improved as it does not efficiently support the current requirements imposed by intelligent manufacturing control and automation systems, namely in terms of flexibility, expandability, agility, and reconfigurability, since they present a weak response to changed production styles and highly dynamic variations (Marrón, Minder, and Karnouskos 2012).

Thus, a migration from traditional production systems characterized by vertical applications, centralized approach and rigidity to an agile plug-and-produce system that is dynamically adaptable to changing production environment open to new features and functions, flexible to different processing tasks, and modular to enable quick and economical changes is needed.

The solution might see companies redesign new manufacturing systems that not only produce higher-quality products at lower costs, but are also able to quickly and effectively respond to abrupt changes in their environment, enabling production to continue working despite the failure of a single component (Chalfoun et al. 2014; Harrison and Colombo 2005) and combining plug-and-produce devices in order to achieve a flexible manufacturing environment based on rapid and seamless reconfiguration of machinery and robots in response to operational or business events (Boschi et al. 2016).

1.1.2 Industrial Challenges

Combining these different aspects, it is quite obvious that today manufacturing domain is facing several challenges that are radically disrupting the existing competition and value-creation rules (Brand 2016).

First of all, the *lack of widely accepted standard* can be considered as one of the main barriers for the easy introduction and deployment of the new production systems. In fact, there is currently a plethora of different realizations of the same basic concepts developed by research projects and local industrial organizations alike, plug-and-produce lacks common reference architecture and standardized interfaces able to prevent the formation of larger user communities and to reduce the integration effort. This is not a technological challenge or even a conceptual challenge anymore but a question of consolidation, harmonization, and ultimately standardization.

Second, *maintaining high levels of quality over small lot sizes* can be considered as one the most relevant challenges that the manufacturing sector has to face. In fact, any changeover between product variants even if it appears to be very little is likely to require a retuning of the production system (re-ramp-up) to ensure that the quality objectives can be consistently achieved. Engineering out any changeover variation currently is achieved with sophisticated tools and fixtures that are prohibitively expensive. This restricts the ability of Original Equipment Manufacturers (OEMs) to reduce their lot sizes or adds substantial costs to the components. Soft-methods are required that can help to ensure first time right even after production system changes.

Third is the *increasing speed of change*. In a truly global market, the speed of change is increasing while at the same time it is becoming increasingly harder to predict. Forecasting is becoming increasingly difficult and economic risk of long manufacturing lead times resulting from inflexible production assets, high effort of production changes over and long distance from the market is becoming increasingly unsustainable. Europe's competitive advantage has to come from highly flexible resources with even higher levels of productivity.

Moreover, the *resistance of change* can have a high impact on adopting this new manufacturing concept. In fact, existing skill sets and engineering methods that have been used for many years will either become obsolete or have to be substantially adjusted and converted to new engineering approach of automation components, machines, and systems able to facilitate an advanced optimization and dynamic decision-making models, and therefore able to overcome a lot of resistance on all levels of the business and technology supply chain. Furthermore, the demand for highly skilled workers able to operate in the Industry 4.0 factory is increasing and, as a consequence, the need for educated, flexible, and knowledge-based workforce has to be supported by a coherent set of tools and methodologies able to sustain the creation, development, and management of advanced skills at all the levels of the company (Colombo, Schleuter, and Kircher 2015; De Carolis, Taisch, and Tavola 2017).

In this context, another obstacle could come from *the endorsement and missing acceptance of the new more digitalized and flexible production system* in the industrial domain. Doubts on the robustness of the approach, the possibility to integrate old and new environments and issues related to human factors as well as a lack of communication could generate skepticism and eventually stop the evolution. Therefore, specific attention needs to be paid to involvement, motivation, and communication to all the stakeholders. Acceptance should not only address technological issues but also target economic aspects because the missing verification of economic feasibility might be another obstacle for the acceptance.

Fourth is the *legacy production equipment and systems*. In several highly industrialized countries it is possible to find many production assets that have been grown and refined over many generations and through several industrial revolutions. While this clearly is a strength point, it is also becoming an inhibitor for change. The main risk today is that change will be forced on local industry from other more agile economies that can result in sudden and difficult to manage changes. Hence, an approach has to be adapted for enabling a well-managed transformation of existing legacy equipment and systems (Schulte and Colombo 2017).

Fifth, a *large number of diverse interest groups are operating on the stage.* The automation technology, system integrator, and end user community in different industrial domains are notoriously diverse and often lack means to organize themselves. Focus on diverse niche areas and fierce competition make it difficult to agree on wider standards. Strong, influential organizations will have to lead the way that can be pushed down the technology supply chain in one direction but is also open for all actors to participate and strive within their respective areas (EU HORIZON2020 FoF PERFoRM 2015-2018).

And last, the *techno-economic risk of new paradigm adoption* is too high or at least is not a well-identified reliable approach to forecast real benefits out of large investments. The manufacturing industry is traditionally very careful adopting new technologies especially as a more fundamental change of not only an individual focused area but also the engineering and even a business approach is involved. It will take a consolidated effort of key representative industrial organizations to make the first step to show the benefit for the wider industrial automation community. It is the challenge of the first adopter of technology. Everyone is waiting for someone else to make the first step at the moment. Once the benefits have been clearly demonstrated in a real life industrial and business environment, others will quickly follow.

1.2 Industrial Automation Current Approaches

1.2.1 Technology Perspective: Current Solutions

This section introduces an overview of recent technology developments that support the implementation of tools and paradigm for the deployment of CPS in the industrial manufacturing automation domain facilitating the realization of true plug-and-produce and shop floor agility.

For years several emergent engineering approaches and related information-communication-control technologies (ICCT), such as multi-agent systems (MAS), SOA, plug-and-produce systems, cloud and fog technologies, Smart Big Data and Analytics, among others, have been researched and prototype developed in a variety of research and innovation activities (see, e.g. Colombo et al. 2014; EU FP6 SOCRADES Consortium 2009). The confluence of these results with the latest developments in digitalized mechatronics, industrial cyber-physical systems (ICPS), systems-of-systems engineering, IoT, Industry 4.0, and Internet of Services (IoS) is opening a new broad spectrum of innovation possibilities for practitioners and industrialists. The

PERFoRM approach deals with this confluence, and this book allows introducing it to the broad scientific and industrial community.

1.2.1.1 Cyber-Physical Systems in Manufacturing

The term "cyber-physical systems" was coined in the United States of America in 2006, with the realization of the increasing importance of the interaction between interconnected computing systems and the physical world. The introduction of CPS in the industry appeared as an unavoidable process, supported by international initiatives and standardization bodies (see, e.g. ARTEMIS Industry Association 2016; Colombo et al. 2016; CPS and IoT Platform in Japan 2015; EU-CPSOS 2018; EU-US ICT Collaboration 2018; National Science Foundation 2018; SPARC 2015).

Multiple definitions of CPS are provided in literature. The early definition by Lee (2008) defines CPS as integration of computational and physical processes. Embedded computers and networks monitor and control the physical processes, usually with feedback loops where physical processes affect computations and vice versa. Wang, Törngren, and Onori (2015) reformulated this definition saying that CPS use computations and communication deeply embedded in and interacting with physical processes, so as to add new capabilities to physical systems. The authors recommend to get more specific information about a concept map of CPS, as they are public proposed by the Department of Electrical Engineering and Computer Sciences at University of California, Berkeley (Ptolemy project 2018).

Monostori et al. (2016) highlight that most of the researchers point to the origins of CPS to embedded systems, which are defined as a computer system within some mechanical or electrical system meant to perform dedicated specific functions with real-time computing constraints. These embedded systems are characterized by tight integration and coordination between computation and physical processes. According to this conception, in CPS, various embedded devices are networked to sense, monitor, and actuate physical elements in the real world. Moreover, the interaction between the physical and the cyber element is of key importance. In fact, CPS is about the intersection, not the union, of the physical and the cyber. It is not sufficient to separately understand the physical components and the computational components. We must understand their interaction. CPS, when applied in production, rely on the latest and foreseeable further developments of computer science, information and communication technologies, and manufacturing science and technology (Colombo et al. 2016; Monostori et al. 2016).

Specific for manufacturing, integration is the key that can be facilitated by CPS. Manufacturing involves multi-sector activities with quite a broad range of stakeholders. The challenge in manufacturing is the integration of the equipment such that all levels of production may communicate, and CPS shows the promise of potential applications in manufacturing (Wang et al. 2015).

Through the ability to interact and expansion capabilities of the physical world using computing power, communication technologies, and control mechanisms, CPS allow feedback loops, improving production processes and optimum support of people in their decision-making processes. By using the corresponding sensor technology, CPS are able to receive direct physical data and convert them into digital signals. They can share this information and access the available data that connects it to digital networks, thereby forming an IoT. CPS are undoubtedly the integration with embedded systems and systems in real time. This kind of integration is teamed with multitude of tools and systems, such as engineering systems based on knowledge, artificial intelligence, and existing installed systems, which are all together transformed into a new system called CPS (Colombo et al. 2014; Colombo et al. 2017; Hozdić 2015).

A flow of information exists between the physical and the virtual world (Figure 1.1).

Based on the definitions identified in literature, the following comprehensive definition for CPS in manufacturing is proposed:

CPS are autonomous collaborating entities able to sense, and act as physical objects (embedded systems), as well as communicate on global networks, compute, and store data and information in the cyber world through their computerized companion (digital twin) (De Carolis, Taisch, and Tavola 2017).

Their key characteristics are as follows (Carolis, Taisch, and Tavola 2017):

- The overall life cycle approach, exploiting CPS both within and outside the shop floor, with a pervasive data storing, organizing, and sharing with different entities in different phases.
- The ecosystem approach that will derive from the full interoperability of systems based on shared vision and standards and their ability to defy rigid standardized hierarchies but create dynamic structures from their articulated functions.

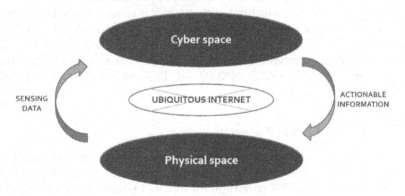

FIGURE 1.1
The flow of information between the physical and the cyber world. (From Hozdić 2015.)

- The new added value services for customers and operators, creating new value for companies, operators, and customers, from the utilization, production, and design processes via aggregation of multiple specific application components.

CPS are clearly perceived as the pivotal enabler for a new era of real-time Internet-based communication and collaboration among value-chain participants; for example, devices, systems, organizations, and humans. When the CPS penetrate the industrial settings, the way enterprises conduct their business is immediately revolutionized from a holistic viewpoint, that is, from shop floor (Operational Technology (OT)) to business (Information Technology (IT)) interactions, from suppliers to customers of tangible and intangible (services) products, and from design to support across the whole product and data and service life cycle (Colombo et al. 2017).

ICPS blur the fabric of cyber (including business) and physical worlds and kick-start an era of system-wide collaboration and (digitalized) data- and information-driven interactions among all stakeholders of the value chain (Colombo et al. 2017). In this sense, from the PERFoRM approach perspective, embedded networked sensors allow the connection and interaction of the factory's physical elements, enabling the exchange of information in real time. Data of interest can be acquired and analyzed, a feedback can be provided, and consequent decisions for running optimized business applications can be actuated.

Overall, the technological basis for enabling conventional production equipment to become cyber-physical production resources today is available in principle (EU FP6 SOCRADES Consortium 2009); research and innovation here focus more on improvements and miniaturization of information technology components. Of major concern here is the development of standards for achieving an easier integration of different components, as well as providing required methodological support for planning and realizing the enablement of existing production equipment (De Carolis 2017).

Regarding the available technological basis, it is needed to analyze the concept and the current scenario of IoT/Cooperating Objects, agent-based CPS, plug-and-produce devices using SOAs and Web services, and cloud-based CPS in industrial automation facilitating production and energy optimization (Bosch 2018; EU HORIZON 2020 FoF PERFoRM Deliverable 1.1 2016; Schneider Electric 2018).

1.2.1.2 Internet of Things/Cooperating Objects

This section introduces an overview of recent technology developments that support the implementation of Cooperating Objects Network of Excellence (CONET) based on the IoT paradigm for the deployment of CPS in the manufacturing automation domain (see, e.g. Marrón, Minder, and Karnouskos 2012; Karnouskos and Colombo 2011).

One can consider that "Cooperating Objects are modular systems of autonomous, heterogeneous devices pursuing a common goal by cooperation in computations and in sensing and/or actuating with the environment" (Marrón, Minder, and Karnouskos 2012). The core idea behind amalgamating the physical and virtual (business) world is to seamlessly gather useful information about objects of the physical world and use the information in various applications in order to provide some added value. A lot of applications can be put under the umbrella of IoT. The close interaction of the business and real world will be achieved by auxiliary services provided in a timely fashion from networked embedded devices.

These will facilitate the collaboration not only among them but also with online services that will enhance their own functionality. As explained in the book by Marrón, Minder, and Karnouskos (2012) there are several areas that share common ground with Cooperating Objects; however, what differentiates them is the mix of the degree of physical and feature elements that creates the right recipe for a specific area. For instance, Cooperating Objects focus mostly on the cooperation aspects while considering the rest of them only as enabling factors to achieve cooperation. Other approaches, for example, IoT, focus mostly on the interaction and integration part while cooperation is optional. So the differentiating factor among all areas is not the distinct characteristics but which of them they employ (depending on the scenario) and at what degree.

1.2.1.3 Agent-Based CPS

The MAS approach (Li and Mehnen 2013; Parsons and Wooldridge 2002) constitutes a software engineering paradigm that offers an alternative to design decision-making systems based on the decentralization of functions over a set of distributed entities. The MAS paradigm replaces the centralized control by a distributed functioning where the interactions among individuals lead to the emergence of "intelligent" global behavior (Colombo, Schoop, and Neubert 2006). This allows reaching a high degree of autonomy and cooperation, without a fixed client-server structure, and consequently to provide the development of systems exhibiting modularity, flexibility, robustness, and reconfigurability. CPS based on agents are being applied to manufacturing and automation domains, as illustrated in several review surveys (Leitão 2009; Marik and McFarlane 2005; Parsons and Wooldridge 2002).In these systems, the complex systems are based on distributed, intelligent, adaptive nodes, the agents, with the overall control system emerging from the interaction among these individual behaviors. Agent-based CPS solutions have also been applied to other domains, such as power energy systems, as illustrated in the article by (Leitão, Mařík, and Vrba 2013). The existing road blockers associated with the agent-based CPS (Parsons and Wooldridge 2002) are usually associated with the lack of interoperability, the lack of development methodologies, and the weak adoption of agents technologies by industry. In fact, in spite of the potential benefits of the MAS technology, currently, there is a gap between the development in laboratorial and simulated environments and the deployment in industrial environments, since the deployed agent-based solutions are few and mainly running in laboratorial environments (Parsons and Wooldridge 2002), with few exceptions; for example, the industrial multi agent-based platform "FactoryBroker™" solution implemented at the Mercedes-Benz engine plant in Stuttgart-Untertürkheim to develop a new kind of production system for manufacturing cylinder heads, reported, for example, in articles by Colombo, Neubert, and Schoop 2001 and Schoop et al. 2002.

1.2.1.4 Plug-and-Produce Devices Using Service-Oriented Architectures and Web Services

This section introduces an overview of recent technology developments that support the implementation and application of SOA-based CPS in the manufacturing automation domain, which is a highly relevant topic for the agile plug-and-produce approach promoted by PERFoRM (see, e.g. Colombo and Stamatis 1998; Colombo et al. 2014).

The recent trends in the technology developments associated with automation devices facilitate the digitalization: Web service protocols are now embedded into a chip, integrated into industrial automation and control devices. Typical production equipment, such as transport units, robots, and also sensors and valves, are considered as modules integrating mechanic, electronic, communication and information processing capabilities. This means that the functionalities of the modules are exposed via Web services into a network. Embedding Web service protocols into the automation device; for example, Devices Profile for Web Services (DPWS) or Object Linking and Embedding (OLE) for Process Control-Unified Architecture (OPC-UA) allows transforming the traditional industrial equipment into a CPS-node of an information communication network. This node will be able to expose and also to consume "Services." Moreover, depending on the position and interrelation of a node to other nodes of the network, it could be necessary to compose, orchestrate, or choreograph services.

The digitalization (i.e. the cyber part) of the mechatronic module (i.e. the physical part) transforms it into a unit that is able to "Collaborate" with other units. That is, a module that communicates via an industrial Internet network with others, exposing or consuming "Services" related to automation and control functions (IoS). Different specifications and functionalities of a collaborative mechatronics module and the corresponding smart automation device are digitalized and the resulted "Services," classified according to the position of the specification and/or function within the product or process life cycle, can be offered to the network for being consumed by other nodes (digitalized mechatronics modules, production components and systems.

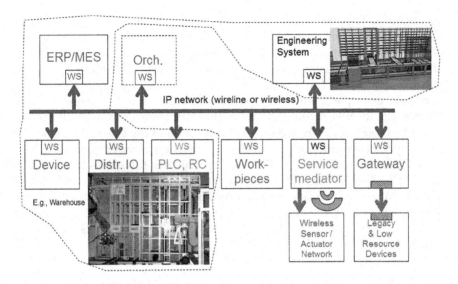

FIGURE 1.2
SOA (Web service–based) shop floor (logic architecture). (From Candido et al. 2009.)

The block with the denomination "Orch" shown in Figure 1.2 represents a module that is able to compose and orchestrate "Services." This logic function can be implemented in a centralized or distributed manner, depending on the kind of digitalized and virtualized system. This means, orchestration engines can be deployed into one or more smart automation devices, that is, another Software (SW) component and processing engine inside of the smart device. Devices are "motors," "valves," "conveyors," "storages," "Human-Machine-Interface (HMI)," "drives," etc., that is mechatronic components with an embedded Central Processing Unit (CPU) and consequently embedded Information-Communication-Technology (ICT) capabilities, that is digitalized-data and -information processing, offered by DPWS, OPC-UA, or similar Web service SW-stack. In this sense, Programable Logic Controller (PLC) and robot controllers (RC) can also be digitalized and transformed into "service producer/consumer" integrating, for example Web service capabilities.

One of the major outcomes of the Web service–based virtualization of a shop floor is the possibility to manage the whole behavior of the plant by the interaction of Web services, that is, exposition, consumption, orchestration, choreography, and composition of the different kinds of services exposed by the different SOA-compliant smart devices and systems. A deeper analysis of the SOA-based automation systems allows us to confirm that the SOA-based virtualization, applied to an enterprise, makes a clear transformation of the traditional hierarchical ISA´95-compliant (ISA Organization 2018b) or the Purdue Enterprise Reference Architecture (PERA)-compliant enterprise architecture (PERA 2018) into a "logical" flat architecture. This major and fundamental outcome of the Web service–based virtualization of a shop floor relies on the fact that the "Services," when they are exposed using the same protocol, for example, DPWS, OPCUA, etc., are directly consumed, composed, and/or orchestrated in an independent way of the source (where these services are physically originated). For example, a Web service exposed by the MES (Mendes et al. 2008) component (located in the ISA'95 Level 3) can immediately be composed with a Web service generated by a Valve (located in the ISA'95 Level 1). Web service protocols embedded into the automation devices such as DPWS and OPC-UA (Candido et al. 2009; SIRENA 2005; SOCRADES 2009; EU FP6 SOCRADES Consortium 2009) are now solutions provided by major device manufacturers. In addition, new technologies have been identified as candidates to develop real-time scalable industrial SOA-based applications, for example, DPWS/Efficient XML Interchange (EXI) and also as combining with MAS technology (Huhns 2002).

1.2.1.5 Cloud-Based CPS in Industrial Automation Facilitating Production & Energy Optimization

This section introduces an overview of recent technology developments that support the implementation of cloud-based CPS in the manufacturing automation domain.

Several projects were and are directly or indirectly investigating cloud-based aspects of manufacturing sector (see, e.g. EU FP7 AESOP Consortium 2010–2014). Activities from standardization bodies are described in the book by Bettenhausen and Kowalewski (2013) as well as in a recent book by Colombo, et al. (2014). For example, Adolphs and Epple (2015) put cloud-based CPS as follows: "Global networks, which are the primary pillars of the modern manufacturing industry and supply chains, can only cope with the new challenges, requirements and demands when supported by new computing and Internet-based technologies."

The fusion of CPS and cloud constitutes the "Cloud of Things" (Acatech—National Academy of Science and Engineering), which flourishes based on services offered to devices and systems, as well as depends on data from devices and intelligence built on the interaction among the physical and cyber (virtual) world. The benefit of utilizing the Cloud of Things is that additional capabilities potentially not available in resource constraint devices can now be fully utilized taking advantage of cloud characteristics such as virtualization, scalability, multi-tenancy, performance, life cycle management, etc. The manufacturer for instance can use such cloud-based services in order to monitor the status and energy demand of the deployed machinery and robots, make software upgrades to the firmware of the devices, detect potential failures and notify the user, schedule proactive maintenance and energy shutdown plans, get better insights on the (energy) usage of the appliance and enhance the product, etc. The user, for example, may benefit from introducing new powerful algorithms, executed in the cloud, into plant operation and its optimization.

1.2.2 Scientific Perspective: Current Research Activities

In this chapter, an overview of the state of the art about the methodologies developed in previous research projects is presented. In fact, previous European projects proved different individual solutions necessary to create an agile production system but none of them proved that concept in an integrated form, impeding the progress toward flexibility and agility. One of the aims of PERFoRM is to integrate and harmonize these concepts and to prepare standards to facilitate industrialization and dissemination of plug-and-produce devices, reusing the acquired knowledge on the described technologies.

A summary of the latest state-of-the-art standards and technologies in manufacturing automation systems proved by former European Union founded projects in the last 15 years, used as knowledge background for initiating the PERFoRM approach, is given in Table 1.1.

1.2.3 Governmental Perspective: Current Initiatives

The needs of improving the industrial sector and to transforming it into a new concept based on digital paradigm, is an emerging global need. Moreover, as there is a strong impact of digital technologies on industry, many G20 states have launched their own initiatives. Digitalization will not only give rise to new business models and new prospects, but will also entail certain challenges. There is a broad consensus that stakeholders from industry, science, politics, labor representatives, and society need to work together to master the challenges and to reap the benefits of the fourth industrial revolution. The dialogue in the G20 is of key importance and can support the process of deepening a global understanding and exchanging best practices. In fact, countries worldwide are trying to strengthen their manufacturing industries by setting up programs, networks and initiatives. During the "Digitising Manufacturing in the G20-Initiatives, Best Practices, and Policy Approaches" in Berlin in March 2017, representatives from eight national initiatives and the EU discussed the national focus that they have initiated in order to foster the development of Industry 4.0 in their country and where they see the potential for international cooperation (Digitising Manufacturing in the G20 2017).

The discussion revealed that opportunities to develop new expertise and smart solutions within a globally interconnected system stand side by side with concerns about not being able to keep pace with the rapid developments in today's world.

TABLE 1.1

State-of-the-Art Standards and Technologies in Automation Systems (European Spectrum Overview)

Project	Objective	Technologies Involved	Solutions and Results
GRACE inteGration of pRocess and quAlity Control using multi-agEnt technology	Developing, implementing and validating a cooperative multi-agent system MAS which operates at all stages of a production line, integrating process control with quality control at local and global levels.	The GRACE architecture consists of a MAS that, as a Middleware, allows storing, analyzing and exchanging data between the Manufacturing Execution System (MES) and the production line control (standard PLCs) (GRACE Consortium 2013).	The project has developed a multi-agent architecture designed for a factory where production processes are subject to planned changes of set-point, typical of on-demand production systems, and to a large variety of disturbances and changes in process parameters and variables. This approach is in line with the current trend to build modular, intelligent, and distributed control systems, called decentralized manufacturing system (DMS).
IDEAS Instantly Deployable Evolvable Assembly Systems	Implementing of agent technology on commercially available control boards.	The IDEAS architecture is based on the MASCOTT tool implementation. This tool is built on existing state of the art description tools enhancing it with the mechatronic concepts. The tool provides the capability (Skills) of each equipment module to be present in a given system. This information is crucial for the creation of the mechatronic agents, for the involvement of Agent Configurator Tool and for the deployment of Resource Agents and Transportation System Agents, which are able to execute the defined skills (IDEAS Consortium 2013).	The project developed a fully distributed and pluggable mechatronic environment capable to self-organize itself and control at the shop floor level. The obtained results are the integration of different modules at the shop floor, the pluggability in runtime, and the distributed diagnosis. Each module is responsible for diagnosing itself and the entire system is capable of checking the propagation of problems and readapt whenever a component/module is plugged without requiring programming effort in order to manage unpredicted behaviors.

Project			
IMC-AESOP ArchitecturE for Service-Oriented Process-Monitoring and - Control	• Proposing a system-of-systems approach for distributed dynamically collaborative monitoring and control based on service-oriented architecture (SOA). • Building a foundation for predictive performance of such SOA architecture based on a formal approach to event-based systems.	The cloud-based architecture was built as a SOA, which connected field devices with higher-level systems, specified as cyber-physical systems (CPS), focusing on SCADA functionalities (but also allowing MES and Enterprise Resource Planning (ERP). This architecture has been implemented in energy management, process industry, district heating, and engineering processes. Assets such as robots, storage, paper machines, lubrication systems, inspection, and maintenance were considered as cyber-physical components and systems (EU FP7 AESOP Consortium 2010–2014).	IMC-AESOP prototyped and implemented in different industrial use cases a cloud-based SOA-based SCADA/DCS infrastructure applying CPS technology for enabling cross-layer service-oriented collaboration, i.e. not only at the horizontal level, e.g. among cooperating devices and systems but also at the vertical level between systems located at different levels of an enterprise architecture.
Self-Learning project Reliable Self-Learning production systems based on context-aware service	Developing highly reliable and secure service-based self-learning production systems aiming at merging the world of secondary processes (e.g. maintenance, energy efficiency, scheduling) with the world of control by using context awareness and data mining techniques.	In self-learning production system, service oriented integration is adapted at the device level to support the application from the upper level. This vertical collaboration between the device level SOA and enterprise layer are perceived using Middleware technologies in Web services platform (Self-Learning Consortium 2009-2013).	The project allowed the creation of three components: • Context Extractor, that is in charge with detection and interpretation of data from existing database systems, data servers, and file systems; • The Adapter, that is in charge with real-time adjustments of control and maintenance parameters, generation of maintenance plans, execution and identification of new parameters to be considered in the control loop; • The Learning Module to learn relying on data mining and operator's feedback to update execution of adaptation and context extraction at runtime.

(Continued)

TABLE 1.1 (Continued)

State-of-the-Art Standards and Technologies in Automation Systems (European Spectrum Overview)

Project	Objective	Technologies Involved	Solutions and Results
SOCRADES project - Service-oriented cross-layer infrastructure for distributed smart embedded systems	Developing and implementing in industrial manufacturing environments a design, execution and management platform for next-generation industrial automation systems, exploiting SOA paradigm both at the device and at the application level. Road mapping the penetration of the SOA paradigm and associated technologies into the industrial environment, showing with TRL6-7 the applicability in different industrial manufacturing systems.	SOCRADES focused its efforts on the device level addressing problems concerning wireless sensor/actuators networks (WSAN) in order to develop reliable and efficient communication carriers. The Enterprise Integration (ERP/MES) is important to obtain and exploit benefits coming out from the use of SOA as Middleware, and also system engineering and management (Engineering System) are relevant for the dynamic reconfigurability and re-configuration (EU FP6 SOCRADES Consortium 2009).	Execution and management of platform for next-generation industrial automation systems, exploiting the SOA paradigm both at the device and at the application level focusing on four main technology areas: • Ad hoc networking service platform—SOA; • Wireless sensor/actuator networking infrastructure; • Enterprise integration based on the SOA paradigm; • System engineering and management to automatically generate control logic software and to allow the support of distributed control system configurations.
PRIME Plug-and-produce intelligent multi-agent environment based on standard technology	Developing a multi-agent architecture using plug-and-produce principles for configuring production systems through innovative human-machine interaction (HMI) mechanisms	The PRIME approach is based on standard technologies and languages for the integration and networking of heterogeneous control systems from different equipment suppliers inside a production line (PRIME Consortium 2015.	The project allowed the realization of: • software toolbox for integrating and enhancing machines and production systems with plug-and-produce agent capabilities and interfaces for seamless integration; • methodology for system behavior modeling and real-time awareness to support system evolution linked to process performance and product volume variability; and • multi-agent control architecture for module integration including legacy equipment.

FLEXA advanced FLEXible Automation cell	Creating the tools, methods, and technologies needed to define and validate an automated flexible cell that can manufacture a generic process chain allowing for safe human interaction and deliver quality assured parts for the European aerospace industry.	The FLEXA Cell Controller uses an architecture that connects to ERP using Middleware/MES, database, simulation and planning, and the connection to automation resources (assets) (FLEXA Consortium 2014).	The project allowed the realization and demonstration of a flexible cell controller, fully functional control software, in two cells. It also developed an integrated system from planning to production and a common interface for communication between automation resources.
ReBorn— Innovative Reuse of modular knowledge- Based devices and technologies for Old, Renewed and New factories	Demonstrating strategies and technologies that support a new paradigm for the reuse of existing production equipment in factories. It aims at delivering models and methods for innovative factory layout design with modular plug-and-produce equipment and flexible low-cost mechanical systems	High focus on the development of hardware (modular reconfigurable assembly equipment, electric presses, AM cells for automatic production of specific tooling, and highly flexible resistance welding cells) (ReBORN Consortium 2016).	Providing a significant step toward 100% reuse of equipment focusing its approach on three main areas: Modular plug-and-produce equipment, in-line adaptive manufacturing; innovative factory layout design techniques and adaptive (re) configuration; flexible and low-cost mechanical systems for fast and easy assembly and disassembly.
EMC2-Factory— Eco Manufactured transportation means from Clean and Competitive Factory	Developing a radically new paradigm for cost-effective, highly productive, energy-efficient, and sustainable factories.	From the architectural view, scheduling (ERP), SCADA, simulation, data analytics, and motion control have been considered (EMC2-Factory Consortium 2014).	Development of procedures for systematic identification of improvement measures for energy efficiency in factories (direct production processes, shop floor, environment) and several methods and tools for energetic improvement (production and factory planning, scheduling, process chain planning).

(Continued)

TABLE 1.1 (Continued)

State-of-the-Art Standards and Technologies in Automation Systems (European Spectrum Overview)

Project	Objective	Technologies Involved	Solutions and Results
ARUM—Adaptive Production Management	Improving planning and control systems for complex, small-lot products manufacturing in order to overcome the increasing risks from product immaturity, the complex and highly customized products for small production series and the weak integration of engineering to production (horizontal) and enterprise ICT to shop floor automation (vertical).	The project proposed a solution that includes: • intelligent Enterprise Service Bus (iESB); • operational MAS schedulers, combining agent-based technology with mathematical solvers; • a set of ontologies, to support the data representation and exchange between the different set of tools and to allow the different tools to have a transparent access to the ARUM data; • a set of ESB enhancements, namely life cycle management tool, node manager, transformation service; and • event analysis tool, enabling the adjustment of the manufacturing process parameters (ARUM Consortium 2015).	Improvement of the reactivity during ramp-up time in small-lot complex manufacturing; a better information flow in the manufacturing shop floor; a more accurate and responsive scheduling and strategic planning tools; and seamless tools integration by the use of the iESB backbone.
MANUCLOUD—Distributed Cloud product specification and supply chain manufacturing execution infrastructure	Development of a service-oriented IT environment as a basis for the next level of manufacturing networks by enabling production-related inter-enterprise integration down to the shop floor level.	Implementation and the creation of self-descriptions during the production equipment engineering, e.g. with PLC programming (MANUCLOUD Consortium 2014).	The solution consists of flexible composition of production facilities (throughout production sites) for simplified production network composition for personalized products, integration of production facilities by means of self-descriptions of "manufacturing services," and platform for manufacturing service management.

Two observations describing the global phenomenon of digital production are figured out:

1. The transformation process takes different forms, but the challenges are similar. While Spain's focus lies on extending its IT competencies, Germany and Italy are banking on their strengths in the automation and manufacturing sectors. All the initiatives' representatives have common concerns regarding cyber security, changing workforce requirements, regulatory frameworks, and standards.

2. The transformation stresses the need for and benefits of new multilateral cooperation. It is important that many initiatives set up networking opportunities and foster an exchange of best practices, including with international partners, carrying out cross-sectoral and cross-country business alliances, and setting up digital innovation hubs and research and cluster networks that are among national support strategies. As a matter of fact and looking into the European industrial manufacturing landscape, this approach is in line with European activities that aimed to encourage the EU Member States to start their own initiatives adapting the EU approach to national and regional specificities, in order to share experiences, collaborate even further when action at EU-level is needed, and to add significant value (Digitising Manufacturing in the G20 2017).

Following these considerations, key recommendations and areas of action have been identified:

- Encourage the creation or further development of national initiatives involving relevant stakeholders on digitizing manufacturing and support learning partnerships between them. These initiatives can contribute to the digitalization of economies and facilitate international cooperation.
- Sharing best practices is essential to facilitate digital transformation on a global level and to increase transparency about the ongoing activities in the G20 countries.
- To follow the path to digital transformation and recognize and grasp the opportunities presented by Industry 4.0 small- and medium-sized enterprises (SMEs) need particular support.
- Centers of excellence, digital hubs (for start-ups), and clusters should be encouraged in order to inform and support them, as well as to explore synergies with various partners, including large companies. For SMEs, easy access to test beds is necessary and there is a need for testing facilities set up by various international partners.

- The development of international standards should be industry and market led, based on principles of openness and transparency. Consensus-based standardization and de facto standardization should be included.
- Work within international forums and consortia should be intensified in order to ensure interoperability in Industry 4.0.
- Companies should be helped to identify IoT risks, to consider a consistent, effective, and resilient design to protect security-/safety-critical assets and maintain a safe and secure state, and to dispatch and share information.
- Digital skills and competencies are driving forces for innovation and competitiveness in G20 and partner countries' economies. Digital literacy and digital skills should be elements of all forms of education and professional training throughout people's lives. Starting from early education to vocational and university education to lifelong learning—the acquisition of digital skills is essential in all these periods, especially in the transition from job to job.
- Governments now have a window of opportunity, and should create an environment for innovation that includes incentives for investments in new technologies.
- Incentives for international research collaboration and new business-to-business connections should be fostered (Digitising Manufacturing in the G20 2017).

In this context, it is important to promote common, consensus-based standardization activities to support thriving economies on the scenario of the EU digital single market strategy. Shared characteristics of digital manufacturing related standards are interoperability, openness, scalability, plug-and-play mechanisms, and security, which support a seamless and easy integration of different IoT solutions.

For these reasons, France, Germany, and Italy as important players in the field of digitization in Europe have started initiatives to keep up and improve their position in the manufacturing industry. Alliance Industrie du Futur in France, Plattform Industrie 4.0 in Germany, and Piano Industria 4.0 in Italy have agreed to join forces working on a shared action plan toward internationalization as end-to-end digital continuity and global standardization are of crucial importance for a digitized economy.

The aim of the trilateral cooperation between Platform Industrie 4.0, Alliance Industrie du Futur, and Piano Impresa 4.0 is to press ahead jointly with existing digitalization processes in the spirit of the European ideal. This trilateral cooperation is focused on three core topics, and they in turn will be dealt with in three joint working groups.

First working group will identify relevant standards, coordinate efforts made toward standardization, work on the harmonization of an administration shell, and find ways of integrating SMEs in the field of standardization. The group will benefit here from the achievements of the Franco-German working group.

The second group aims to promote SME integration and test beds. In order to make digitalization more accessible for SMEs, this group collects and combines example applications from all three countries, disseminate and complete Industrie 4.0 scenarios, and promote an international network of test infrastructures.

The third working group has the objective to provide a political support. It exchanges ideas about best practices from political solutions and programs from many different sectors and levels, coordinating common positions of the three countries at European level and international forums. For these reasons, this group has initial cooperation talks between the three national initiatives, which were held in Berlin in March 2017 at the high-level conference "Digitising Manufacturing in the G20." Since then, a joint plan of action has been developed, in which various measures and scheduled results have been named. On June 20, 2017, the plan was adopted by the Board of Cooperation at a first joint meeting in Turin, Italy. Experts from industry, science, and politics from all three countries are now continuing work steadily on the three core topics (Platform I 4.0 2018).

The three initiatives in France/Germany/Italy

About Alliance Industrie du Futur Created in July 2015, *(Platform Industrie 4.0 and Industrie du Futur 2018)—Alliance Industrie du Futur brings together professional organizations from industry and digital technology along with academic and technological partners, all rallying around a common ambition: To make France a leader in the world's industrial renewal. Alliance Industrie du Futur is tasked with putting into practice the National Industry of the Future project launched by the French Government in April 2015, in the framework of the organizational overhaul of the New Face of Industry in France. Alliance Industrie du Futur has a particular mandate to support companies in the transformation of their business models, their organization, and their design and marketing methods, in a world in which new tools based, for example, on the digital, on additive manufacturing, on new materials, and on advanced robotics are bringing down the barriers between industry and services.*

About Platform Industrie 4.0 *(Platform i40 2018)—Platform Industrie 4.0 is the central alliance to coordinate the shaping of the digital structural shift of German industry. It brings together all those who are shaping Industrie 4.0 and bundles the forces and know-how of a diverse range of actors—be they companies, associations, trade unions, science, or politics. With over 300 participants from over 150 organizations the Platform is one of the largest international and national networks, it supports German companies—particularly small and medium-sized companies—in implementing Industrie 4.0. It offers decisive input by providing practical solutions and examples of company practices from across Germany, concrete recommendations for action and test environments.*

About Piano Nazionale Industria 4.0—*In order to boost productivity and accelerate technological upgrading, the National plan "Industria 4.0" of the Italian*

Government is focusing on strategic measures to support innovative investments and empower skills, such as: Super and Hyper amortization schemes, tax credits on R&D and on profits from intangible and patented assets, strengthening of vocational training, creation of I4.0 Technological Clusters and Industrial PhDs. Moreover, complementary measures being implemented are the establishment of Competence Centers and of a network of Digital Innovation Hubs, Ultra Broadband with a fibre to the factory approach, cooperation on IoT open standards and interoperability, easier access to financing and productivity salary taxation exchange (Platform Industrie 4.0 2016)

1.3 PERFoRM Ambitions

It is necessary to bring manufacturing industry a next generation of agile manufacturing systems that are dynamically reconfigurable and evolvable to enable evolution, self-organization and adaptation along the system life cycle, facing the challenges of continuously and rapidly changing market conditions and increasingly smaller lot sizes, and shorter lead time and time-to-market requirements. These systems should be based on modular plug-and-produce components (with built-in intelligence) with all the different actors involved in the manufacturing system life cycle (module suppliers, system integrators, end users, etc.) brought together to smoothly design, deploy, ramp-up, operate, and reconfigure the new generation of production systems.

In response to the previous mentioned challenges, PERFoRM aim can be decomposed into the following main objectives:

1. *Development of a modular and agile manufacturing control system,* based on plug-and-produce system concepts, to include the different control layers from Level 1 to Level 3 of the ISA 95 (ISA Organization 2018a) (covering the distributed intelligence at device and system level). This integrated (holistic) approach, covering the different manufacturing system layers, will enhance the achievement of self-adaptation, reconfigurability, and evolution. Important aspects to consider here are the definition of a network of plug-and-produce components and the dynamic composition and aggregation of these components to face the dynamic condition change.

2. *Integration and testing of dynamic, robust and flexible local and global monitoring and optimization algorithms,* implementing feedback control loops to allow the dynamic and robust online monitoring of key performance indicators (KPIs) that enables the rapid intervention at the shop floor level by means of augmented reality solutions (e.g. real-time adjustment of the process or the adjustment of

parameters reflected in the product quality). The system combines local and global levels in a natural manner; for instance, plug-and-produce devices may support local adaptation that can be merged into global system adaptation. Energy reduction strategies will be transformed into commands that will be sent back to the resources to control energy consumption.

3. *Harmonization and standardization of methods that enable the plug-and-produce readiness* (in terms of machines, processes, etc.). The design and development of standard interfaces, including industrial automation Middleware, to link the cyber physical artefacts that compose the plug-and-produce components (as a system of systems), complemented with parameterizing methodologies to build these systems, will accelerate the adoption of plug-and-produce systems into industrial environments. A very important element for achieving agile production systems is a reliable and adaptable communication infrastructure (industrial Middleware) to support the vertical integration plug-and-produce production resources with higher-level applications. This Middleware/infrastructure will offer seamless access to a range of software services for the plug in of ready components/resources and legacy components/resources as well as other lower level services, such as simulation services, visualization services, and communication services.

4. *Implementation and demonstration of methods, methodologies, strategies for transforming existing production systems into plug-and-produce production systems,* establishing guidelines for a smooth migration from a traditional system to agile plug-and-produce systems in a secure and efficient way through the use of plug-and-produce device adapters. Successful guidelines for this migration, similar to those addressed in (Jammes et al. 2012), are crucial since the approach presented here corresponds to an entirely new way of thinking at shop floor and at enterprise management levels. Experience has demonstrated that technical solutions by themselves are not sufficient to ensure successful adoption by industry. In this way, it is fundamental to create a plethora of assisted procedures that guide system integrators in creating true plug-and-produce systems that can be rapidly changed over.

These objectives require research activities to implement a solid manufacturing Middleware based on encapsulation of production resources and assets according to existing paradigms (e.g. CPS, service-based architectures, cloud services, etc.), the development of advanced and modular global monitoring and optimization algorithms for reconfiguration of machinery, robots, and processes and, finally, to ensure the full interoperability, the harmonization and standardization of methods and protocols to enable the plug-and-produce readiness in heterogeneous environments.

Hence, the vision of the PERFoRM approach is not to start from scratch but to focus on overcoming the last remaining barriers to starting exploiting the knowledge already generated by these former FP7 projects. Although the projects were successful on an individual basis, the vision of truly agile production systems based on rapidly integrable plug-and-produce components has not yet been achieved. It is the right time now to harmonize the outputs of the large body of work create in this area and take the next step toward standardization and industrial exploitation. This should be underpinned by large-scale industrial demonstrations and comprehensive awareness raising activities to ensure uptake beyond the project consortium.

Achieving real industrial and business impacts requires a careful consideration of migration of legacy environments to the new approach and, where standard devices and/or applications remain, the coexistence and integration of existing systems.

Focusing on the technology perspective, the project contributes to advance the technological readiness in the agent-based plug-and-produce systems in several aspects. In fact, first of all, a control layer architecture will be designed (adapting from the results of previous projects) to support the development of integrated solutions for the fast reconfigurable individual machines and robots, optimizing the changeover times and costs. Second, the lack of methodologies and the weak adoption by industry will be covered through the development of migration methods that will allow the use of the proposed principles by system integrators, module and equipment developers, and end users.

Applying the PERFoRM approach, an industrial Middleware component based on cloud-based CPS, which is ready to be applied in real industrial applications, has been specified, developed, and deployed in different industrial use cases. PERFoRM adopts the intelligent adapter approach and combines it with key performance metrics and industrial network profiles for energy usage, and integrates them into plug-and-produce adapter and legacy system integration concept. The focus is then put on harmonizing performance capture and communication protocols and models including energy usage information. This is also linked to the simulation and optimization framework to achieve more agile and self-adaptive production systems.

One of the major results of applying the approach is then the development of an innovative digitalized and harmonized production system, which is able to combine the flexibility and the good utilization ratio of a job-shop organization with the efficiency (especially small stocks) of a line production through innovative decentralized control routines and reconfigurable production equipment. Production cells can be aligned and connected in multiple ways intelligently guiding products from one production step to another. Products will "know" their own specific way and their respective requirements in regard to production equipment, which is enabled through adequate technology and interfaces.

Following certain decision algorithms (criteria might be "travel time to cell", "time till availability of production cell" etc.), products can find their next production cell respectively and their way through the production system. These highly complex production systems cannot be framed with a straightforward simulation method anymore. Instead, the design is an iterative process that relies on intermediate results of simulation runs that are capable of reflecting the dynamic system behavior (nonlinear or stochastic material flow).

The innovation work behind the application of the PERFoRM approach strongly focuses not only on applicability and innovative results, but especially on migration strategies and ways to transform production lines to digitalized harmonized production systems. Migrating a factory into a plug-and-produce–based production environment displays an expansive paradigm shift concerning the design of production systems and applied communication and decision structures. Achieving this transformation in one single step cannot be expected when migrating an existing factory. The migration process has to be conducted in a series of consecutive steps, running through a series of intermediate states, each needing to be economically feasible for itself.

1.4 PERFoRM Approach

1.4.1 PERFoRM Objectives

The fulfillment of the objectives addressed in the previous section can be achieved by using a common, but customized for different industrial domains, plug-and-produce approach based on digitalization and CPS technologies that are mature enough, that is, have the adequate technology readiness level for industrial deployment (see, e.g. European Union 2014; NASA 2012).

For this reason, the main relevant PERFoRM objectives are to consolidate the main development that are affecting the current scenario in terms of industrial, technological, and research perspectives and to integrate them into a common reference architecture for automation and control that can be deployed into existing production environments. In this regard, PERFoRM aims to foster the generation of a new manufacturing environment characterized by enhanced perception, autonomy, and intelligence of manufacturing assets (machineries, robots, and workers), which are able to simultaneously guarantee the dynamic reconfiguration of manufacturing processes, while assuring efficiency and effectiveness of the system in terms of performance and safety.

PERFoRM is seen to be a key enabler to embrace the conceptual TRANSFORMATION to the penetration of the CPS paradigm into the industry, aiming to deploy it into running-living production environments, implementing an EVOLUTIONARY concept encompassing the shop floor side, and also the dynamic business-sensitive composition of applications starting from standard building blocks.

1.4.2 PERFoRM Project Structure

Core of the approach is to establish an adequate Middleware component, which is able to link industrial field devices (OT) with systems of the upper IT levels, in order to migrate the existing production systems based on the traditional centralized, vertical, and rigid paradigm into a new distributed paradigm based on plug-and-produce–enabled production resources.

In this context PERFoRM leads to develop a next generation of agile manufacturing systems that are dynamically reconfigurable and evolvable to cope with smaller lot sizes and shorter lead time and time to market, based on plug-and-produce systems concept.

The applicability of the approach has been validated in four industrial test beds belonging to different industrial domains, product sizes, production volumes, and process types, to ensure a broad and sound validation of the concept and the platform behind.

1. Siemens AG—Compressors (see Chapter 10)
2. IFEVS—Micro electrical vehicles (see Chapter 12)
3. Whirlpool—Home appliances (see Chapter 11)
4. GKN—Aerospace (see Chapter 13)

The PERFoRM industrial use cases could take several advantages in utilizing flexible and reconfigurable systems based on CPS philosophy, although they have diverse goals due to the different nature of their products. They allow the validation of the PERFoRM approach in four different sectors.

1.5 Conclusions

PERFoRM is dealing with the digitalization, harmonization, prototype implementation, and industrial proof of different technologies to develop a solution suitable for the current and future industrial manufacturing ecosystem, under the perspective of the fourth industrial revolution. In fact, on the one hand the proposed solution integrates different technological aspects provided by the new digital emerging technologies; on the other hand, it tries to respond to one of the needs characterizing the current market trend: The demand for flexibility and reconfigurability of the production system. In this manner, the industrial production systems are able to respond to mass and extreme customization of tangible products and services. In particular, the application of the PERFoRM approach aims at making reality an advanced ICPS architecture composed of digitalized industrial plug-and-produce components at both IT and OT levels of an enterprise.

This chapter presented an overview of the knowledge background behind the PERFoRM approach, reporting on some of the latest developments and achievements on several fronts, that is, the architecture, the design, the enabling technologies, and a representative set of applications of digitalization methods and tools as well as CPS introduced and used in the industrial manufacturing environment. Moreover, in order to show how many cross-cutting challenges, addressed in the latest literature (see, e.g. Colombo et al. 2017; Foehr et al. 2017) are being addressed by PERFoRM, a positioning of the approach in the international arena, both from strategical and scientific and technological perspectives, has been done.

References

Adolphs, P., and U. Epple. 2015. "Statusreport: Referenzarchitekturmodell Industrie 4.0 (RAMI4.0)." April 0.

ARC Advisory Group. 2018. "Collaborative Manufacturing Management (CMM)." Accessed October 12, 2018. https://www.arcweb.com/industry-concepts/collaborative-management-model-cmm.

ARTEMIS Industry Association. 2016. "Strategic Research Agenda SRA of the ARTEMIS Industry Association." Accessed October 12, 2018. https://artemis-ia.eu/documents.html.

ARUM Consortium. 2015. "ARUM." Accessed October 15, 2018. https://cordis.europa.eu/project/rcn/104761_de.html.

Bettenhausen, K, and S. Kowalewski. 2013. "Cyber-Physical Systems: Chancen und Nutzen aus Sicht der Automation." April.

Bi, Z. M., S. Y. T. Lang, W. Shen, and L. Wang. 2008. "Reconfigurable manufacturing systems: The state of the art." *International Journal of Production Research*, Vol. 46 (4): 967–92.

Robert Bosch Manufacturing Solutions GmbH 2018. "Connected Industry." Accessed October 15, 2018. https://www.bosch-connected-industry.com/.

Boschi, F., C. Zanetti, G. Tavola, and M. Taisch. "Functional Requirements for Reconfigurable and Flexible Cyber-Physical System." In *IECON 2016 - 42nd Annual Conference of the IEEE Industrial Electronics Society*, pp. 5717–5722.

Brand, D. 2016. *MESA MOM Capability Maturity Model Version 1.0.*

Cala, A., M. Foehr, D. Rohrmus, N. Weinert, O. Meyer, M. Taisch, et al. 2016. "Towards industrial exploitation of innovative and harmonized production systems."

Candido, G., F. Jammes, J. Barata, and Armando W. Colombo (2009). "Generic Management Services for DPWS-Enabled Devices." In *IECON 2009 - 35th Annual Conference of IEEE Industrial Electronics (IECON 2009)*, pp. 3931–3936.

Chalfoun, I., K. Kouiss, N. Bouton, and P. Ray. 2014. "Specification of a reconfigurable and agile manufacturing system (RAMS)." *Int. J. Mech. Eng. Autom.* 1 (6): 387–94.

Colombo, A. W., R. Neubert, and R. Schoop. 2001. "A Solution to Holonic Control Systems." In *IEEE International Conference on Emerging Technologies and Factory Automation 2001*, pp. 489–98. Piscataway, NJ: IEEE.

Colombo, A. W., R. Schoop, and R. Neubert. 2006. "An agent-based intelligent control platform for industrial holonic manufacturing systems." *IEEE Trans. Ind. Electron.* 53 (1): 322–37. doi:10.1109/TIE.2005.862210.

Colombo, A., and K. Stamatis. 1998. "Towards the factory of the future: A service-oriented cross-layer infrastructure." *Innov. Technol. Transf.* 1 (98): 16–21.

Colombo, Armando W., Dirk Schleuter, and Matthias Kircher. 2015. "An Approach to Qualify Human Resources Supporting the Migration of SMEs into an Industrie4.0-Compliant Company Infrastructure." In *IECON 2015 - Yokohama: 41st Annual Conference of the IEEE Industrial Electronics Society: November 9-12, 2015, Pacifico Yokohama, Yokohama, Japan*, edited by Kiyoshi Ohishi and Hideki Hashimoto, pp. 3761–3766. Piscataway, NJ: IEEE.

Colombo, Armando W., Stamatis Karnouskos, Okyay Kaynak, Yang Shi, and Shen Yin. 2017. "Industrial cyber-physical systems: A backbone of the fourth industrial revolution." *EEE Ind. Electron. Mag.* 11 (1): 6–16. doi:10.1109/MIE.2017.2648857.

Colombo, Armando W., Stamatis Karnouskos, Yang Shi, Shen Yin, and Okyay Kaynak. 2016. "Industrial cyber-physical systems [scanning the issue]." *Proc. IEEE.* 104 (5): 899–903. doi:10.1109/JPROC.2016.2548318.

Colombo, Armando W., Thomas Bangemann, Stamatis Karnouskos, Jerker Delsing, Petr Stluka, Robert Harrison, et al. 2014. *Industrial Cloud-Based Cyber-Physical Systems.* Cham: Springer International Publishing.

"CONET." www.cooperating-objects.eu.

CPS and IoT Platform in Japan 2015. "CPS and IoT." Accessed October 15, 2018. https://www.jeita.or.jp/cps-e/ (CPS and IoT Platform in Japan, 2015).

"Cyber-Physical Systems Driving force for innovation in mobility, health, energy and production."

De Carolis, A. 2017. "A Methodology to Guide Manufacturing Companies towards Digitalization." Accessed October 15, 2018. https://www.politesi.polimi.it/handle/10589/136972.

De Carolis, A., Marco Taisch, and Giacomo Tavola. 2017. *sCorPiuS:-Future Trends and Research Priorities for CPS in Manufacturing.* Accessed October 15, 2018. https://www.scorpius-project.eu/sites/scorpius.drupal.pulsartecnalia.com/files/documents/sCorPiuS_D3_3_Final%20roadmap_whitepaper_v1.0.pdf.

Delsing, Jerker, Fredrik Rosenqvist, Oscar Carlsson, Armando W. Colombo, and Thomas Bangemann. 2012. "Migration of Industrial Process Control Systems into Service Oriented Architecture." In *IECON 2012: 38th Annual Conference on IEEE Industrial Electronics Society: Montreal, Quebec, Canada, 25-28 October 2012*, pp. 5786–5792. Piscataway, NJ: IEEE.

"Digitising Manufacturing in the G20." 2017.

EMC2-Factory Consortium. 2014. "EMC2-Factory." Accessed October 15, 2018. https://cordis.europa.eu/project/rcn/101388_de.html.

ETSI World Class Standards. 2009. *ICT Shaping the World: A Scientific View.* Chichester: Wiley.

EU-CPSOS, 2018. 2018. "Cyber-Physical Systems of Systems." Accessed October 12, 2018. http://www.cpsos.eu/project/what-are-cyber-physical-systems-of-systems/.

EU FP6 SOCRADES Consortium. 2009. *SOCRADES Introducing a Service Oriented Infrastructure for Industry.* Accessed October 10, 2018. https://www.youtube.com/

watch?v=BCcqb8cumDg, https://www.youtube.com/watch?v=K8OtFD6RLMM, or https://www.youtube.com/watch?v=K8OtFD6RLMM&t=404s.

EU FP7 AESOP Consortium. 2010–2014. "ArchitecturE for Service-Oriented Process-Monitoring and Control (IMC-AESOP)." Accessed October 10, 2018. https://cordis.europa.eu/project/rcn/95545_de.html.

EU FP7 GRACE Consortium 2010-2013. "Integration of Process and Quality Control using Multi-Agent Technology". Accessed October 10, 2018. https://cordis.europa.eu/project/rcn/94796/factsheet/en

EU FP7 IDEAS Consortium 2010-2013. "Instantly Deployable Evolvable Assembly Systems". Accessed October 10, 2018. http://www.ideas-project.eu/index.php/presentation-ideas-project.html

EU FP7 Self Learning Consortium 2009-2013. "Reliable Self-Learning Production Systems Based on Context Aware Services". Accessed October 10, 2018. https://cordis.europa.eu/project/rcn/92738/reporting/en

EU FP7 PRIME Consortium 2012-2015. "Plug and PRoduce Intelligent Multi Agent Environment based on Standard Technology". Accessed October 10, 2018. https://portal.effra.eu/project/1031.

EU HORIZON2020 FoF PERFoRM 2015-2018. Accessed October 10, 2018. https://cordis.europa.eu/project/rcn/198360_de.html.

"EU HORIZON 2020 FoF PERFoRM Deliverable 1.1: Report on decentralized control & Distributed Manufacturing Operation Systems for Flexible and Reconfigurable production environments." 2016. Accessed October 10, 2018. http://www.horizon2020-perform.eu/index.php?action=documents.

European Union, Definitions T.R.L. 2014. "Technology Readiness Levels." Accessed October 15, 2018. https://ec.europa.eu/research/participants/data/ref/h2020/wp/2014_2015/annexes/h2020-wp1415-annex-g-trl_en.pdf.

EU-US ICT Collaboration, IoT/CPS. 2018. "IoT/CPS." Accessed October 15, 2018. http://www.picasso-project.eu/.

FLEXA Consortium. 2014. "FLEXA." Accessed October 15, 2018. https://cordis.europa.eu/result/rcn/59062_en.html.

Foehr, Matthias, Jan Vollmar, Ambra Calà, Paulo Leitão, Stamatis Karnouskos, and Armando W. Colombo. 2017. "Engineering of Next Generation Cyber-Physical Automation System Architectures." In *Multi-Disciplinary Engineering for Cyber-Physical Production Systems: Data Models and Software Solutions for Handling Complex Engineering Projects*. Vol. 8, edited by Stefan Biffl, Arndt Lüder, and Detlef Gerhard, pp. 185–206. Cham: Springer International Publishing.

Harrison, Robert, and Armando W. Colombo. 2005. "Collaborative automation from rigid coupling towards dynamic reconfigurable production systems." *IFAC Proceedings Volumes*. 38 (1): 184–92. doi:10.3182/20050703-6-CZ-1902.01571.

Hozdić, E. 2015. "Smart factory for industry 4.0: A review." *Int. J. Mod. Manuf. Technol.* 7 (1): 28–35.

Huhns, M. N. 2002. "Agents as Web services agents as Web services." *Agens on Web* 6 (August): 93–95.

ISA Organization. 2018a. "International Standard for the Integration of Enterprise and Control Systems: ISA'95." Accessed 10 October 2018. https://www.isa.org/belgium/standards-publications/ISA95/.

ISA Organization. 2018b. "ISA'95." Accessed October 15, 2018. https://www.isa.org/isa95/.

Jammes, Francois, B. Bony, P. Nappey, A. W. Colombo, J. Delsing, J. Eliasson, et al. 2012. "Technologies for SOA-Based Distributed Large Scale Process Monitoring and Control Systems." In *IECON 2012-38th Annual Conference on IEEE Industrial Electronics Society, Quebec, Canada, 25-28 October 2012*, pp. 5799–5804. Piscataway, NJ: IEEE.

Karnouskos, S., and A. W. Colombo. 2011. "Architecting the next generation of service-based SCADA/DCS system of systems." *Appl. Sci.* 312–17.

Lee, E. A. 2008. "Cyber Physical Systems: Design Challenges Oriented Real-Time Distributed Comput. 2008." *In 11th IEEE Int. Symp. on Object and Component-Oriented Real-Time Distributed Comput*, pp. 363–69.

Leitão, P., V. Mařík, and P. Vrba. 2013. "Past, present, and future of industrial agent applications." *IEEE Trans. Ind. Informatics.* 9 (4): 2360–2372.

Leitão, Paulo. 2009. "Agent-based distributed manufacturing control: A state-of-the-art survey." *Engineering Applications of Artificial Intelligence.* 22 (7): 979–91. doi:10.1016/j.engappai.2008.09.005.

Li, Weidong, and Jörn Mehnen, eds. 2013. Cloud Manufacturing: Distributed Computing Technologies for Global and Sustainable Manufacturing. *Springer Series in Advanced Manufacturing.* London: Springer; Imprint: Springer.

Malhotra, V., T. Raj, and A. Arora. 2010. "Excellent techniques of manufacturing systems: RMS and FMS." *Int. J. Eng. Sci. Technol.* 2 (3): 137–42.

MANUCLOUD Consortium. 2014. "MANUCLOUD." Accessed October 15, 2018. https://cordis.europa.eu/result/rcn/59193_en.html.

Marik, V., and D. McFarlane. 2005. "Industrial adoption of agent-based technologies." *IEEE Intell. Syst.* 20 (1): 27–35. doi:10.1109/MIS.2005.11.

Marrón, Pedro J., Daniel Minder, and Stamatis Karnouskos. 2012. The Emerging Domain of Cooperating Objects: Definitions and Concepts. *Springer Briefs in Electrical and Computer Engineering.* Berlin, Heidelberg: Springer Berlin Heidelberg. http://dx.doi.org/10.1007/978-3-642-28469-4.

McKinsey & Company. 2015. "Industry 4.0 How to Navigate Digitization of the Manufacturing Sector." Accessed October 10, 2018. http://worldmobilityleadershipforum.com/wp-content/uploads/2016/06/Industry-4.0-McKinsey-report.pdf.

Mendes, J. M., Paulo Leitão, Armando W. Colombo, and Francisco Restivo (2008). "Service-Oriented Control Architecture for Reconfigurable Production Systems." In *2008 6th IEEE International Conference on Industrial Informatics (INDIN)*, pp. 744–49.

Monostori, L., B. Kádár, T. Bauernhansl, S. Kondoh, S. Kumara, G. Reinhart, et al. 2016. "Cyber-physical systems in manufacturing." *CIRP.* 65 (2): 621–41.

NASA, T. R.L. 2012. "Technology Readiness Level." Accessed October 15, 2018. https://www.nasa.gov/directorates/heo/scan/engineering/technology/txt_accordion1.html.

National Science Foundation, C. P.S. 2018. "Cyber-Physical Systems: Enabling a Smart and Connected World: NSF-CPS 2018." Accessed October 15, 2018. https://www.nsf.gov/news/special_reports/cyber-physical/.

Parsons, Simon, and Michael Wooldridge. 2002. "An introduction to multi-agent systems." *Autonomous Agents and Multi-Agent Systems.* 5 (3): 243–54. doi: 10.1023/A:1015575522401.

PERA. 2018. "PERA Enterprise Architecture and Life Cycle Mode." Accessed October 15, 2018. http://www.pera.net/.

Platform i40, 2018. "Industrie4.0." Accessed October 12, 2018. https://www.plattform-i40.de/I40/Navigation/DE/Home/home.html.

Platform Industrie4.0 and Industrie du Futur. 2018. "Plattform Industrie 4.0 and the France's Alliance Industrie du Futur." Accessed October 12, 2018. https://www.plattform-i40.de/I40/Redaktion/EN/Standardartikel/international-cooperation-industrie-du-futur.html.

"Ptolemy project 2018." Accessed October 20, 2018. https://ptolemy.berkeley.edu/projects/cps/.

ReBORN Consortium. 2016. "ReBORN." Accessed October 15, 2018. https://cordis.europa.eu/result/rcn/194626_en.html.

Schneider Electric. 2018. "EcoStruxure™ Platform." Accessed October 15, 2018. https://www.schneider-electric.de/de/work/campaign/innovation/platform.jsp?account=44354&&account=44354& (Schneider Electric "EcoStruXure Platform").

Schoop, R., A. W. Colombo, B. Suessmann, and R. Neubert. 2002. "Industrial Experiences, Trends and Future Requirements on Agent-Based Intelligent Automation." In *IECON-2002: Proceedings of the 2002 28th Annual Conference of the IEEE Industrial Electronics Society: Sevilla, Spain, November 5-8, 2002*, pp. 2978–2983. Piscataway, NJ: IEEE.

Schulte, Daniel, and Armando W. Colombo. 2017. "RAMI 4.0 Based Digitalization of an Industrial Plate Extruder System: Technical and Infrastructural Challenges." In *Proceedings IECON 2017 - 43nd Annual Conference of the IEEE Industrial Electronics Society: China National Convention Center, Beijing, China, 29 October - 01 November, 2017*, pp. 3506–3511. Piscataway, NJ: IEEE.

Platform Industrie 4.0 and Industrie du Futur 2018. Accessed on October 10, 2018. https://www.plattform-i40.de/PI40/Redaktion/DE/Downloads/Publikation/plattform-i40-und-industrie-du-futur-scenarios.html

"SIRENA 2005." Accessed October 15, 2018. http://www.sirena-itea.org/.

"SOCRADES 2009." Accessed October 15, 2018. http://www.socrades.eu/.

SPARC, Robotics i. E. 2015. "Robotics 2020, Multi-Annual Roadmap." Accessed October 12, 2018. https://www.eu-robotics.net/cms/upload/downloads/ppp-documents/Multi-Annual_Roadmap2020_ICT-24_Rev_B_full.pdf.

Wang, Lihui, Martin Törngren, and Mauro Onori. 2015. "Current status and advancement of cyber-physical systems in manufacturing." *Journal of Manufacturing Systems*. 37: 517–27. doi:10.1016/j.jmsy.2015.04.008.

Wang, Zhaohui, Houbing Song, David W. Watkins, Keat G. Ong, Pengfei Xue, Qing Yang, et al. 2015. "Cyber-physical systems for water sustainability: Challenges and opportunities." *IEEE Commun. Mag.* 53 (5): 216–22. doi:10.1109/MCOM.2015.7105668.

2

Technologies and Standards

Olga Meyer, Daniel Schel, Daniel Stock, Greg Rauhöft

(Fraunhofer Institute for Manufacturing Engineering and Automation (IPA))

Jeffrey Wermann, Steffen Raasch, Armando W. Colombo
(Institute for Informatics, Automation and Robotics (I²AR))

CONTENTS

2.1 Technologies

Innovation in the development of smart reconfigurable production systems (Koren et al. 2013) is promoted by standards and specifications of the new technologies. A large number of these technologies must be realized within a factory to enable Industry 4.0 (I4.0)-compliant transformation. Recent developments show that the I4.0-enabling technologies (Wan, Cai, and Zhou 2015), in particular Internet of Things (IoT) (IIoT 2018; Xu, He, and Li 2014), Cyber-Physical Systems (CPSs) (Colombo et al. 2014; Monostori 2014) and Industrial Cyber-Physical Systems (ICPSs) (Colombo et al. 2017), are already contributing to a higher flexibility, reconfiguration, and adaptability of production systems in real production environments. Numerous industrial products, including smart manufacturing systems and other digital devices, are already on the market. However, the majority of presented solutions must be replaced by open and standardized solutions because these still focus on vendor-specific scenarios and bring severe limitations in their deployment.

2.1.1 Interoperability and Communication

For a better understanding of the Production harmonizEd Reconfiguration of Flexible Robots and Machinery (PERFoRM) framework, this section summarizes the previous work and presents detailed information of the supportive technologies.

Interoperability and communication are crucial criteria while constructing a robust and efficient technology backbone for flexible and reconfigurable production solutions. These criteria are often understood as a synonym for the Internetof Things. In fact, the IoT has been defined in the Recommendation International Telecommunication Union (ITU)-T Y.2060 as a global infrastructure for the information society, enabling advanced services by interconnecting (physical and virtual) things based on existing and evolving interoperable information and communication technologies (ITU-T, Telecommunication Standardization Sector of ITU 2013). Applied in the industrial field, the IoT represents a concept to establish a seamless connection of the virtual and physical worlds in a manufacturing factory and adopt new forms of communication between manufacturing machines, products, and various production processes (Kortuem et al. 2010; Meyer et al. 2017; Xu, He, and Li 2014). Accordingly, there is a large heterogeneous mix of communication technologies, which need to be adapted in order to meet the PERFoRM requirements for flexibility, reconfigurability, and plug-and-produce capability of the connected systems and applications.

Additionally, interoperability and communication are the key functionalities for the realization of the vertical and horizontal integration and networking of manufacturing or service systems along with the life cycle of the product. Integration requires intelligent cross-linking and digitalization of business processes on different hierarchical levels within the organization and among supply networks. It is, therefore, very important to meet the crucial requirements of connected components and products before these are deployed and used in their operational environment in the organization (Lin et al. 2017). The integration and interoperability between connected manufacturing systems, products, and service systems, both vertical Operational Technology (OT)/Information Technology (IT) and horizontal connectivity (Givehchi et al. 2017), will only succeed with the help of common, uniform Information and Communications Technology (ICT) standards. For their technical description and implementation, a reference architecture model is necessary.

Common reference architecture models consist of stable construction elements that allow cross-layer semantic interoperability in a system architecture. Applied to I4.0 and IoT needs, a proper reference architecture model is an excellent means to assist the task of identifying, classifying, and evaluating existing technologies and standards.

Currently, various reference architecture models are proposed for the IoT construct by standardization organizations and initiatives (Weyrich and Ebert 2016), as for instance:

- Industrial Internet Reference Architecture (IIRA) (IIC 2017) developed by industrial internet consortium (IIC).
- Reference Architecture Model for Industrie 4.0 (RAMI4.0) (IEC PAS 63088:2017 Smart manufacturing—Reference Architecture Model Industrie 4.0 (RAMI4.0); DIN SPEC 91345:2016-04 Reference Architecture Model Industrie 4.0 [RAMI4.0]), developed by Plattform Industrie 4.0 (Platform i40 2018).
- Internet of Things Reference Architecture (IoT RA) (International Organization for Standardization (ISO)/International Electrotechnical Commission [IEC]), currently under development by ISO/IEC JTC1 (International Organization for Standardization 2016).
- Project IoT-A Architectural Reference Model (IoT-A ARM) (Bauer et al. 2014), developed by the IoT-A project partners of the European FP7 Research.
- Industrial Value Chain Reference Architecture (IVRA) (IVI, Industrial Value Chain Initiative 2018), developed by IVI.
- The Internet of Things World Forum Reference Model (Cisco Systems, Inc., Cisco Systems (USA) Pte. Ltd., Cisco Systems International BV Amsterdam 2014), proposed by the IoT World Forum Architecture Committee.

However, not all models are systematically focusing on the I4.0 needs and, therefore, are lacking consistency. On top, there is a strong need in harmonization of diverse representations of the reference architecture models and their conceptual scope. However, several initiatives have been already launched in this direction, for example, the Industrial Internet Consortium and Plattform Industie 4.0 completed standardization activities and mapped IIRA and RAMI4.0. Though there are differences in the representation and focus domains, the mapped reference architecture models contain similar as well as complementary elements (Lin et al. 2017).

2.1.2 Application Setup

The four application scenarios of PERFoRM outline flexible production lines that can be reconfigured and harmonized during runtime to manufacture a new product variant or to benefit from new production capabilities. In order to integrate a set of field devices or production modules into legacy production lines with minimal overhead and, thus, greatly affect the flexibility at the shop floor being ready to changing customer demands, PERFoRM relies on plug-and-produce capabilities of the connected systems. In general, the PERFoRM factory is a flexible and adaptable factory of the future that consists of digitalized and interoperable modular production facilities, which can be easily adapted, extended, reorganized, and reconfigured to address customer needs and further production demands in a harmonized manner.

Though the vision of the I4.0 foresees realization of a large number of new technologies within a factory, a rush development of innovative ICT technologies cannot be quickly transferred to the industry as far as reliable and robust IT security concepts, architecture models, and relative standards are missing. To solve these problems and, thus, support the transformation of industrial production systems toward the next generation of reconfigurable CPS (Cala et al. 2017; Colombo et al. 2016), PERFoRM is pursuing a clear strategy for technology development and adoption (Cala et al. 2016; Meyer et al. 2017). Specifically, PERFoRM adopts and develops a large variety of technologies and standards that originate from other European and non-European projects (Cala et al. 2016; Colombo et al. 2014; EU MANUFUTURE 2013). Furthermore, it aims at creating a consistent combination of the newly developed information and communication technologies with the classic standardized approaches to meet the requirements, that is, flexibility and reconfigurability, of the production systems and applications. The developed solution focuses on new forms of reconfigurable plug-and-play capabilities of the production systems (Leitão et al. 2016) as well as on a seamless reference architecture and integration technologies (Gosewehr et al. 2017).

Furthermore, for successful implementation of the transformation to the flexible and reconfigurable production systems, a mix of fundamental technologies is required to build the sufficient ICT architecture solution. To define and select the best available technologies it is important to make a clear picture of existing standardized solutions and apply efficient techniques, principles, and concrete steps for a technology gap analysis (Meyer et al. 2017). The PERFoRM approach recognized the opportunities and made a profound analysis of more than 200 technologies that were submitted by various experts from more than 15 relevant European projects (Meyer et al. 2017).

2.1.3 PERFoRM Technology Framework

PERFoRM identifies the DIN SPEC 91345:04-2016, Reference Architecture Model for Industrie 4.0 (RAMI4.0) (IEC PAS 63088:2017 Smart manufacturing—Reference Architecture Model Industry 4.0 (RAMI4.0); DIN SPEC 91345:2016-04 Reference Architecture Model Industrie 4.0 [RAMI4.0]) as the most efficient and suitable reference tool for description of the applied technology framework. RAMI4.0 is a three-dimensional (3-D) model that is internationally recognized supporting as well IoT and Internet-of-Services (IoS) paradigms (https://www.beuth. de/de/technische-regel/din-spec-16593-1/287632675), (German Standardization Roadmap 2018). The vertical dimension represents the digitalization aspects from physical and cyber parts of assets/things that compose , for example, a factory, covering from products to the enterprise and supply chain levels, following the enterprise reference architectures specifications IEC 62264 (Nagorny et al. 2012) and IEC 61512 (transversal RAMI4.0 dimension) (IEC 61512 1995–2006). The horizontal axis of the 3-D RAMI4.0 manages the product life cycle and the value stream specifications of those digitalized assets/things, specified according to the standard IEC 62890 (DIN/VDE 2017).

The PERFoRM technology framework includes a mix of applied technologies. It builds on the vertical axes of the RAMI4.0 represented by six various layers (IEC PAS 63088:2017 Smart manufacturing—Reference Architecture Model Industry 4.0 (RAMI4.0); DIN SPEC 91345:2016-04 Reference Architecture Model Industry 4.0 [RAMI4.0]) and includes crosscutting functionalities and technology references (Figure 2.1). The crosscutting functionalities mirror PERFoRM functional requirements, whereas the technology references give a brief overview of the experienced EU projects (bottom of Figure 2.1), which were identified as the potential technology suppliers of innovative and robust I4.0 solutions for the approach needs. The layer structure and applied technologies of the PERFoRM technology framework is described as follows:

The *asset layer* represents the physical world that includes physical components, software applications, documents as well as humans that are involved in production processes. At this layer, physical and digital tools of PERFoRM such as turning lace (provided by SIEMENS), chassis and powertrain testing area (provided by I-FEVS), micro-flow cell processes (provided by GKN), or microwave oven (provided by WHIRPOOL) are present.

The *integration layer* represents integrated physical components and their digital representation. This level includes such technologies as Human-Machine-Interface (HMI) that allows the integration of a human (e.g., process control via HMI device at the GKN Use Case described in Chapter 13 of this book), as well as computer-aided control of the technical process and technology adapters. Additionally, this layer includes generated events from the connected systems that embody altered or reconfigured states of the real world.

The *communication layer* provides standard communication technologies for service and data formatting toward the integration layer. This layer includes various standardized technologies such as open platform communications

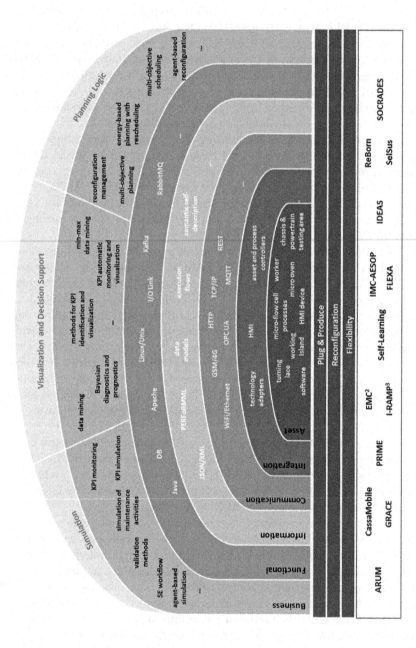

FIGURE 2.1
PERFoRM Technology Framework.

unified architecture (OPC UA), representational state transfer (REST), and message queuing telemetry transport (MQTT) communication protocols (interface technologies of the PERFoRM Middleware); HTTP/TCP, GSM/4G, and Wifi/Ethernet technology.

Figure 2.1 is not exhaustive and depicts only few technologies. The detailed description of this and other technologies and tools will follow in Chapters 5 and 7 of this book.

The *information layer* aggregates the informational representation of data and data models that are generated, preprocessed, used, and modified by the connected physical and software components. Here, the PERFoRM Framework uses the PERFoRM Modelling Language (*PERFoRMML*) that ensures data integrity between the adjoint layers and provides consistent integration of the data among connected components. Thus, the received events are transformed in the universal appropriate format to match the requirements of the PERFoRM Middleware data structure.

The *functional layer* describes technical functionalities of the connected systems. The key technologies of this layer are represented by the PERFoRM Middleware, specified by the Institute for Industrial Informatics, Automation and Robotics (I2AR) of the Hochschule Emden/Leer (HSEL), described in the Chapter 6 of this book, which is a central component of the IT architecture responsible for the horizontal integration of the physical components, on the one hand, and business components, that is, business software applications/services and tools, on the other hand. Furthermore, this layer includes runtime and modelling environment for IT architecture and its back end software applications to support business processes (e.g., *RabbitMQ* an open source message broker software, various data bank management systems for data collection and management, Linux/Unix operating system used for the software stack of the PERFoRM Middleware software). Thus, the functional layer allows remote access to the IT platform and various information and processes of the subsidiary layers. Within the horizontal integration, this layer also ensures seamless integrity of the information flows in the processes and integration of the technical IT systems.

The *business layer* orchestrates various services and application processes that form business logic and ensures integrity of these processes in the value stream. At this layer, PERFoRM integrates several service technologies that provide functional and business support for the PERFoRM use cases. The service portfolio can be divided in three main groups: Simulation services (e.g., agent-based simulation provided by Technical University Braunschweig (TUBS) for the use case of GKN Aerospace (described in Chapter 13 of this book), visualization and decision support service (e.g., key performance indicator (KPI) monitoring with what-if-game functionality provided by the Polytechnic Institut Braganza (IPB) for the use cases at WHIRLPOOL (described in Chapter 11 of this book) and GKN Aerospace (described in Chapter 13 of this book), and planning logic services (e.g., energy-based planning with rescheduling provided by XETICS GmbH for the use case of I-FEVS (described in Chapter 12)).

2.2 Standardization

PERFoRM's standardization strategy aims at identifying and evaluating existing standards that are needed to implement the flexible reconfigurable architecture solution for the next generation of digitalized, reconfigurable, and harmonized production systems. For the application scenarios, worked out within the PERFoRM project scope (EU HORIZON2020 FoF PERFoRM 2015–2018) and described in the following section of the book, the PERFoRM approach identified several standardization gaps and needs.

2.2.1 Overview of the Standardization Landscape

Standardization is the driving force for innovation (Bitkom, VDMA, and ZVEI 2016) and sustainable development. Standards create a secure basis of uniform technical rules that are used to support designers, engineers, operators, and other decision makers (Lu, Morris, and Frechette 2016) in their work both within and between various domains. Within a domain, that is, a specific organization or enterprise, standards provide accurate instructions and lead to cost-efficient realizations of solutions (IERC, European Research Cluster on the Internet of Things January). Due to the focus shift to digital industries, the I4.0 standards have the goal to fill a large niche in the current standardization landscape and support industries in the migration process. Thus, the development of such standards has gained a special priority both at the national and international levels. The development of the standards takes place on different levels (national, regional, and international). Figure 2.2 shows the general overview of the hierarchical structure.

Note: Figure 2.2 is basically a modified and extended version of the picture shown in DIN (2015) from the DIN organization.

National standardization organizations represent the interest at the level of one specific country. Within a country or a territorial division of a country, experts from industry, academic and research bodies, and other interested organizations and associations are involved in the standardization process. The standardization process is open and transparent and usually takes place in standard committees that consist of representatives of all involved parties. The committees are responsible for developing new national, European, and international standards as well as for updating or revising existing standards.

The standardization system at the European level is based on the national pillars, that is, the National Standardization Bodies or the members of the respective standard developing organizations, CEN (European Committee for Standardization), CENELEC (European Committee for Electrotechnical Standardization), and ETSI (European Telecommunications Standards Institute). It is the responsibility of the National Members to implement European standards as national standards.

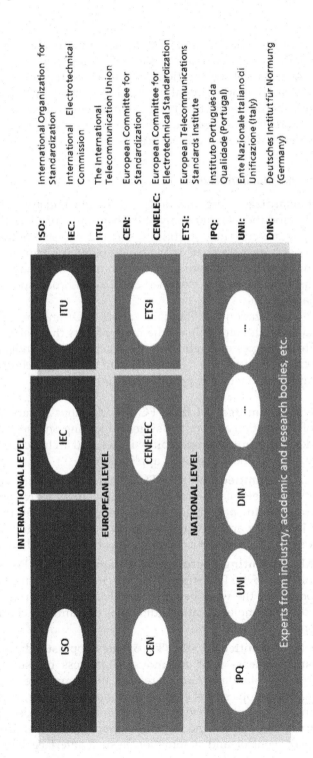

FIGURE 2.2
Standardization at various levels.

The main role of international and European standardization organizations is to overcome technical barriers in standards development caused by differences among technical specifications and standards that are developed independently by each nation, national standards organization, or company. Accordingly, these bodies target to a great extent at harmonization of standards in order to prevent or overcome technical incompatibilities and communication problems of the involved parties.

2.2.1.1 International Standardization

International standardization organizations and other I4.0-relative initiatives in Europe and around the world are only at the beginning of a long journey. It means that they are currently starting to sort out valuable outcomes of the first projects and developments that reveal potential in context of I4.0 standardization. Nevertheless, there are a large number of standardization activities that have been recently launched and have a broad spectrum on various I4.0-related topics.

International Organization for Standardization (ISO) and International Electrotechnical Commission (IEC) are the main players that develop standards for ICT technologies and represent the interest of the industries in the fields of electrical engineering, electronics, and IT. The key contributing technical committees and working groups are as follows:

ISO Technical Committee 184 (ISO/TC 184) develops standards in the field of automation and control systems focusing on integration, physical device control, interoperability issues, and architectures for enterprise systems and automation applications.

ISO Technical Committee 299 (ISO/TC 299) develops international standards in the field of robotics. Robots with a high degree autonomy are expected to open up entirely new opportunities for manufacturing industries and manage dynamic working environments that allow secure human-robot collaboration on the production floor.

ISO Smart Manufacturing Coordinating Committee (ISO/CMCC) is a successor body of the ISO Strategy Groups Industrie 4.0 and has a goal to coordinate the work being undertaken on an international level and develop implementation guidelines that aim at a common international approach.

IEC Technical Committee 65 (IEC/TC 65) develops standards for systems and elements used for industrial-process measurement and control. The standardization areas of the committee involve several important topics of I4.0 and smart manufacturing, such as smart manufacturing framework and system architecture, security for industrial automation and control systems, industrial-process measurement, control and automation system interface between industrial facilities, Industrial communication networks, and life cycle management for industrial systems and products.

IEC System Committee Smart Manufacturing (IEC/SyC) is a newly formed committee that aims to coordinate and harmonize standardization activities and advice the IEC, other standard developing organizations and consortia in the domain of Smart Manufacturing.

ISO/IEC Joint Technical Committee for Information Technology (ISO/IEC JTC 1) is one of the largest technical committees created by the ISO and IEC. The committee develops, maintains, and promotes a wide range of standards in the fields of IT and ICT. Table 2.1 presents a mapping matrix of the JTC 1 activities that are focusing

TABLE 2.1

Relevant Focus Areas of ISO/IEC JTC 1

ISO/IEC JTC 1/	Big Data	Software Applications	(Industrial) IoT	Networking and Middleware	Edge Computing	Cloud Computing	Software Processes and Life Cycle	Smart Systems	Software Interfaces	Security	Identification Technologies
SC 6 Telecommunications and information exchange between systems	x			x		x	x	x			
SC 7 Software and systems engineering		x					x	x			
SC 17 Cards and security devices for personal identification										x	x
SC 22 Programming languages, their environments, and system software interfaces								x	x		
SC 25 Interconnection of information technology equipment		x	x				x	x	x		
SC 27 IT Security techniques	x		x			x	x	x		x	x
SC 31 Automatic identification and data capture techniques	x		x			x		x			
SC 38 Cloud Computing and distributed platforms	x	x	x	x	x	x	x	x		x	
SC 40 IT Service management and IT governance		x					x	x			
SC 41 Internet of Things and related technologies	x	x	x	x	x	x	x	x	x	x	x
SC 42 Artificial intelligence	x	x	x				x	x	x	x	

Note: This table is basically a modified and extended version of the table shown on slide 16 in François Coallier (2018).

on I4.0 topics or can be identified within the scope. To maintain a holistic view of the IoT area and related technologies, JTC 1 established a new entity in the form of a new Subcommittee 41 (SC41). SC41 has already created a substantial portfolio of standards and projects and is currently exploring standardization gaps and needs in the industrial domain.

ISO/IEC Joint Working Group 21 (ISO/IEC JWG 21) was launched in July 2017 with the aim to harmonize the activities and scope between ISO/TC 184 and IEC/TC 65 and liquidate substantial overlaps in the development of Smart Manufacturing reference models, in particular, focusing the life cycle and virtual representation of connected assets.

ISO/IEC Smart Manufacturing Standards Map Joint Working Group (ISO/SMCC–IEC/SEG 7 Task Force) (German Standardization Roadmap 2018) is a joint working group between ISO/SMCC and IEC/SEG and aims at harmonization of I4.0 standards at the international level.

International Telecommunication Union (ITU) is the United Nations' specialized agency for information and communication technologies The **ITU Telecommunication Standardization Sector (ITU-T)** is one of the key sectors in ITU that coordinates standards for telecommunications at international level and commits to the development of recommendations focusing on digital systems and global information infrastructure, Internet protocol (IP) aspects, next-generation networks, Cloud Computing, IoT technologies, and cybersecurity. Recently, ITU-T has also launched activities in blockchain standardization.

European Telecommunication Standards Institute (ETSI) develops globally applicable standards for ICT-enabled systems, applications, and services in various domains. Specifically, in the industrial domain, the TC Smart Machine-to-Machine communications (TC Smart M2M) focuses on interoperability and connectivity requirements and aspects.

2.2.1.2 National Standardization and Key Industrial Initiatives

It is of the great interest of the national standardization organizations and other involved public and private sectors to provide national-level perspectives and actions to European and international standard developing organizations. There are a large number of active national standardization bodies, industrial initiatives, and open platform organizations that are involved in the development of I4.0-related standards and provide efficient technical solutions.

- Among these bodies, Germany takes the leading role with regard to digital innovations in manufacturing technology (German Standardization Roadmap 2018). The main contributors are the

German Institute for Standardization (DIN) (DIN 2018) and the *German Commission for Electrical, Electronic & Information Technologies of DIN and VDE (DKE/VDE)* (DKE/VDE 2018). DIN is an acknowledged national standards body that represents German interests in European and international organizations. DIN is the German member of CEN and ISO. DKE/VDE develops standards and safety regulations for the electrical engineering and IT sectors and represents the German interests in CENELEC and IEC. At the national level, Germany puts strong standardization efforts in the development of the RAMI4.0 model and such topic as administration shells of physical assets, also called I4.0 components. The standards are also published at the international level as IEC PAS 63088 in 2017.

- *National Institute of Standards and Technology (NIST)* (NIST 2018) is a part of the US Department of Commerce and one of the oldest industry's national laboratories. On the national level it has a crosscutting function to bring industrial stakeholders together and provide competence in areas such as smart manufacturing, communications technology and cybersecurity, advanced manufacturing, and disaster resilience. NIST also supports robotics topics in the manufacturing scope as well as the integration of robotics and humans in industrial environment.

- *Ministry of Industry and Information technology of China (MIIT)* is currently promoting a 10-year plan *Made in China 2025* (Wübbeke et al. December 2016), a blueprint for upgrading the country's manufacturing sector and bring the industrial transformation to China. According to this plan, China is going to upgrade the factories in order to become more competitive, innovative, and efficient. Together with the *Standardization Administration of China (SAC)*, MIIT also composed an architecture construction guide for the China Intelligent Manufacturing System Architecture (IMSA) (Li et al. 2018).

- *Industrial Value Chain Initiative (IVI)* (IVI 2018-1) is a Japanese initiative that was commenced in 2015 and is similar to Germany's I4.0 initiative. The Japanese companies have the goal to connect business processes, manufacturing, and information technologies and transform these to the real industrial environment (METI, Ministry of Economy, Trade and Industry May 15, 2015).

- *OneM2M* (oneM2M) is the global standards initiative that covers requirements, architecture, API specifications, security solutions, and interoperability for Machine-to-Machine and IoT technologies. The initiative actively supports industry associations and initiates forums with the focus on specific application and industry requirements and interoperability solutions.

- *Institute of Electrical and Electronics Engineers (IEEE)* (IEEE 2018) develops standards with the main focus on the lower protocol layers, that is, the physical layer. The IEEE *Intercloud Working Group (ICWG)* is creating technical standards for cloud interoperability introducing a federation concept, which provides a dynamic infrastructure and is aimed to support new business models. *The IEEE Industrial Electronics Society (IES) TC on Industrial Agents* is doing significant work in the development and application of industrial agent technology in production, services, and infrastructure sectors. Robustness, scalability, reconfigurability, and flexibility are recognized to be the key benefits of the agent technology when applied to smart industrial systems. The work is carried out in IEEE Standards Association (IEEE-SA) in conjunction with the IEEE Industrial Electronics Society (IES) and the IEEE Systems, Man and Cybernetics Society (SMC). The IEEE P2660.1™ Working Group (WG) (IEEE-SA) is developing recommendations that aim to solve the interface problem when integrating industrial agents with automation control level in the context of industrial cyber-physical systems (ICPSs) (Colombo et al. 2017).
- *Industrial Internet Consortium (IIC)* (IIC 2017, IIC 2018) is developing open interoperability standards and common architectures and is mainly involved in international standardization activities on a collaborative basis.
- *Open Connectivity Foundation (OCF)* (OCF 2018) is an industry group that develops specification standards and provides open source implementations with the focus on interoperability and communications platform.
- *Alliance for Internet of Things Innovation (AIOTI)* (AIOTI, Alliance for Internet of Things Innovation June 30, 2017) was established in 2016 with the support of the European Commission (EC) and is focusing on IoT standardization. In accordance with its current strategy plan, the AIOTI working group on standardization concentrates on activities such as IoT gap analysis, analysis of IoT-related use cases, development of strategies for consolidation of architectural frameworks, reference architectures, recommendations for (semantic) interoperability, security, and personal data protection.
- *Korea Smart Factory Foundation (KOSF)* is a foundation of private companies that has been established under support of Korean government and has the goal to provide sufficient test beds and coordination of smart factory projects.
- *NAMUR (NAMUR)* is an international user association of automation technology in process industries. The association addresses issues that include solutions and systems for the management, steering, control, inspection, and communication of production plants, and other important concept developments and requirements of I4.0 approaches.

- The Society of German Engineers (VDI), the German Electrotechnical Society (VDE), and the Mechanical Engineering and Plant Association (VDMA), *Zentralverband Elektrotechnik- und Elektronikindustrie (ZVEI)*, and Bundesverband Informationswirtschaft, Telekommunikation und neue Medien (BITKOM) are the key contributors to the national standards in Germany that develop standards, that is, with the focus on Reference Architecture Model I4.0, IT security, fundamental concepts for modeling industrial technical assets, their life cycles, and their management in the information world.

- The *Plattform Industrie 4.0* is a joint creation of the three industrial associations BITKOM, VDMA, and ZVEI with the aim to bring together representatives from various sectors in a collaborative process. In the scope of standardization, the main activities focus on the establishment of a common understanding of the I4.0 transformation, in particular, with the focus on industrial sector topics such as employment and training, research and innovation, or security of networked systems.

- *Standardisation Council Industrie 4.0 (SCI 4.0)* (SCI 4.0, 2018) was founded in 2006 by DIN and DKE together with the German industrial associations BITKOM, VDMA, and ZVEI. It initiates standards of digital production in Germany, in fact, the SCI 4.0 takes over the coordination between industry and standardization, that is, mediates between the members of the Plattform Industrie 4.0 and the various standardization organizations. In order to reach the consensual harmonization of I4.0 concepts on a global level, SCI4.0 takes the responsibility for creation and update of the Standardization Roadmap for I4.0 (German Standardization Roadmap 2018).

- *Labs Network Industrie 4.0 (LNI 4.0)* (LNI 4.0, 2018) was set up by companies from the Plattform Industrie 4.0, together with BITKOM, VDMA, and ZVEI and proposes a laboratory and centers to test technical and economic feasibility of new technologies, business models and use cases. LNI 4.0 works in close collaboration with DIN, Plattform Industrie 4.0, and SCI 4.0 to support the standardization process.

2.2.1.3 International Cooperation in Standardization

The international cooperation activities impact the openness and transparency of the standards development process. Such cooperation is expected to strengthen the communication of the standardization organizations and other involved parties and support the development of a robust technological basis for industries. However, there are many barriers that hamper the development and adoption of global standards for industrial digital transformation.

On the one hand, the novelty of I4.0-related topics and diverse technical solutions that emerged in a relatively short period of time on the market lead to a strong fragmentation of the standards landscape. This development resulted in a large variety of overlapping and redundant standards that target at the same topics but propose diverse solutions and are often defined independently in different application sectors. But, as the number of standards produced expands, more opportunity for overlap and redundancy results (Lu, Morris, and Frechette 2016).

On the other hand, the complexity of the traditional vertical silos of the main contributors makes it difficult to establish proper communication at various levels and meet the consistency of terms and concepts (German Standardization Roadmap 2018). At the same time the lack of communication often causes misleading activities that also result in a large number of split focus objectives in standardization and standards organizations often compete with one another to write standards that lead to conflicting standards, defeating the real purpose (Lu, Morris, and Frechette 2016). Therefore, the harmonization actions are particularly important in order to minimize redundant or emerging conflicting standards and achieve consensus among various parties. Therefore, many standards developing organizations and industrial initiatives have already concluded agreements with their international partners in order to ensure cooperation, favorize digitization of industry by developing global standards for I4.0-related topics and, thus, reach harmonization between the national, European, and international levels.

- Within Europe, France, Germany, and Italy have launched several cooperations to support the digitalization process in the manufacturing sector and improve their position in industry. The members of **a trilateral cooperation, Alliance Industrie du Futur in France, Plattform Industrie 4.0 in Germany and Piano Impresa 4.0 in Italy**, are currently working on a shared action plan toward internationalization and global standardization bringing smart manufacturing to the next level. In the Paris Declaration for Smart Manufacturing (Plattform Industrie 4.0, Standardization Council Industrie 4.0, Alliance Industrie du Futur, Ministero dello Sviluppo Economico, UNI -Ente Nazionale Italiano di Unificazione 2018) the platform describes main objectives of the common strategy on international standardization in field of the IoT/I4.0 and shares important issues recognized about smart manufacturing.

- **Germany and Japan** have been advancing active joint discussions through holding repeated meetings of experts from both countries since 2016. The Plattform Industrie 4.0 and the Japanese Robot Revolution Initiative have signed a **joint agreement on their future cooperation** (Plattform Industrie 4.0, Robot Revolution Initiative, Standardization Council Industrie 4.0 2018).

- **The Industrial Internet Consortium (IIC) and the Japanese Industrial Value Chain Initiative (IVI)** (IIC-IVI 2017) established new partnership to speed "Industrial IoT" solutions and have signed an MoU to work together to align efforts to maximize interoperability, portability, security, and privacy in this sector. Joint activities will include sharing of use cases, sharing of information regarding Industrial IoT architecture and best practices for manufacturing, and identifying future joint test bed projects (Industrial Internet Consortium April 26, 2017).

- In July 2017, **IEE-SA announced its collaboration with NIST** aiming to meet the growing demand for standards that address interoperability and standardization of Intercloud Cloud Computing (IEEE Standards Association July 25, 2017).

- **The Industrial Internet Consortium (IIC) and** oneM2M are collaborating to accelerate the availability of different integration technologies and processes, helping to drive standards development and delivering insights from the test beds into specific industry requirements Ome (oneM2M February 8, 2018).

- In November 2017, **CEN and CENELEC** created a joint Focus Groups on Blockchain & Distributed Ledge technologies (DLT) (CEN CENELEC December 14, 2017) to identify specific European standardization gaps and needs for the blockchain technology, focusing on sharing data and managing transactions in a controlled manner, interoperability issues, privacy, security, and European use cases. The group wants to map the identified gas with the current work items in ISO/TC and encourage other industrial initiatives and industrial partner to participate in the international standardization process.

2.2.2 Standardization Gaps and Needs

As part of the PERFoRM approach, several standardization gaps and needs have been identified. They are grouped in several I4.0-related priority areas. The following sections give detailed descriptions of each of those areas.

2.2.2.1 Reference Architectures and Migration Strategies

This area focuses on gaps and needs with regard to recent developments of I4.0-related reference architectures, and I4.0 migration strategies.

Use cases as well as reference architectures present practical systematic scenarios that usually serve as a basis for functional and architectural requirements for interoperable technical systems and server to identify potential technologies and methods. Use cases represent building blocks of a reference architecture within a special context, such as IIoT or value chain.

Several organizations are already contributing to the development of use cases and reference architectures, as already described in the previous sections. Nevertheless, there are some important requirements that should be closely observed while developing a uniform concept for industrial reference architectures, such as consideration of value-added processes in relation to the entire value-added chain (German Standardization Roadmap 2016) as well as sharing applications between multiple companies, integration of technologies through value networks and crosscutting networking, consideration of product-machine communication (Kagermann, Wahlster, and Helbig 2013).

On the other hand, reference architecture models create conditions and requirements for successful implementation of highly flexible concepts that permit a step-by-step migration for an organization from the current technological state to that of I4.0. Furthermore, the number of practical and substantial use cases can significantly contribute to the seamless development of related reference architectures. Therefore, it is important to collect I4.0-related applications in order to identify standardization needs and essential technologies.

As such, to support the standardization process within this area the following gaps should be taken into consideration:

- Minimization of the number of standards involved and development of a uniform structure for use case description.
- Harmonization of use cases developed by various standardization organizations and initiatives and diversity reduction.
- Harmonization of reference architectures and reduction of possible overlaps.
- Consideration of special industrial requirements, in particular aiming at flexibility and reconfiguration.
- Contribution to the development of I4.0 migration strategies and conception of relevant recommendations.
- Use of the standard terminology to avoid misleading descriptions.

2.2.2.2 System Characteristics

The following group focuses on general CPS's requirements and structural representation of digitalized, reconfigurable, and harmonized production systems.

Involved in the product life cycle, a CPS accumulates a large number of important information. Unfortunately, depending on a system, it is usually difficult to extract this information and map to a relevant life cycle phase. Thus, specifications are required that describe concepts and requirements of a CPS on how to preserve the relevant information and, thus, record a life cycle and collect information of the product payload, that is, digital threads and life cycle of digital data (see, e.g., McKinsey & Company 2015).There is already a profound work in this area describing the Industrie 4.0 Component (I4.0 Component) and its Asset Administration Shell (AAS) functions (German Standardization Roadmap 2016). According to the last reported results, the I4.0 Component describes the properties of Cyber-Physical components and systems that are networked in the production environment with others (assets or virtual objects and processes). Figure 2.3 depicts how the I4.0-compliant communication is realized in RAMI4.0 on the basis of the I4.0 Component.

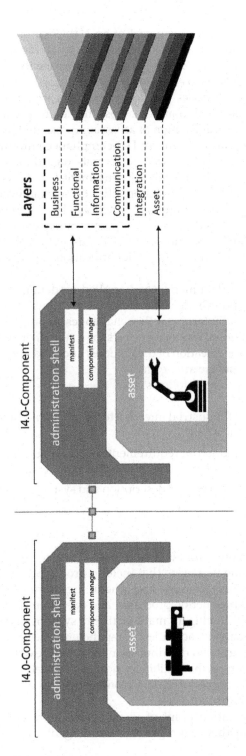

FIGURE 2.3
The RAMI4.0 and the Industrie 4.0 Component (From Fraunhofer IPA, based on VDI/VDE Society Measurement and Automatic Control 2016.)

The AAS as well as an asset is a part of an I4.0-Component (Figure 2.3). An asset is a physical or virtual object that has a value for an organization. It is also considered to be one of the important components of the RAMI4.0. The main task of the AAS is to expose the data and functionality of all assets that are relevant in a life cycle—including products and entire production systems—in a standardized manner across the corresponding RAMI4.0 levels. AAS contains all information of an asset represented by the information model and serves mechanisms for I4.0-compliant communication based on I4.0-defined semantic rules and data formats regardless of the deployment possibilities (VDI/VDE Society Measurement and Automatic Control 2016). The AAS wraps the asset functionalities on several levels, that is, the communication layer, information layer, functional layer, and business layer.

Moreover, in order to participate in a conversation with other systems and applications as an individual unit, a CPS, as self-organizing and self-reconfigurable systems, must have identification and self-descriptive capabilities.

The heterogeneity of data models and applied modeling tools, caused by the lack of proper standards in this area, leads to inconsistency of system's information models and interoperability challenges among various systems. In order to avoid additional modeling efforts for a system or system interface technologies, a clear specification of data models and modeling tools that can be applied for industrial use are needed.

Other identified gaps are as follows:

- Harmonization of industrial interface technologies at the integration and communication levels.
- Contribution to industrial requirements of a CPS, such as reconfiguration, availability, adaptability, scalability, maintenance.
- Focus on a universe semantic description and system aspects of a CPS.

2.2.2.3 Industrial Agents

Agent-based applications provide a suitable solution for flexible plug-and-produce manufacturing scenarios that include production operations such as manufacturing process planning and scheduling, monitoring and control of physical assets and production processes, and operation and management. Applied to distributed intelligent manufacturing systems, industrial agents encapsulate manufacturing activities, wrap heterogeneous software and hardware systems in an open distributed environment, extending the traditional approach that limits the flexibility and reconfigurability of the manufacturing systems (Schoop et al. 2002; Colombo, Schoop, and Neubert 2006; Leitão, Colombo, and Restivo 2006).

From the implementation point of view, agents represent physical and virtual resources and their aggregations. Thus, agents represent not only machines, robots, products, cells, manufacturing applications, and tools but also human operators who cooperate with manufacturing resources through the user interfaces. Considering their peculiar features, that is, provision and support of flexibility, adaptability, and distributed control functionalities, industrial agents play an important role in PERFoRM scenarios.

One of the main standardization needs in this area is the integration of industrial agents with other I4.0-related technologies, in particular with a service-oriented architecture (SOA), to efficiently support the distributed approach scenarios as well as the dynamic composition of manufacturing service applications (Herrera et al. 2008;, Ribeiro, Barata, and Colombo 2008; Nagorny, Colombo, and Schmidtmann 2012). Some of the prototype implementations that based on the use of Multi-Agent Systems (MAS) and SOAs have been already covered in various European projects (Leitão, Colombo, and Karnouskos 2016). However, to achieve the standardization of the industrial agent technology, the industrial agents must demonstrate reasonable capabilities to support service-oriented principles and become conform to the advanced standardized reference architectures. Therefore, PERFoRM investigated the deployment possibilities of industrial agents in the I4.0-related reference architectures, in particular in RAMI4.0.

The properties of an I4.0 Component are similar to the concept of an industrial agent (Figure 2.4). An agent is defined as a "component, that represents physical or logical objects in the system, capable to act in order to achieve its goals, and being able to interact with other agents, when it doesn't possess knowledge and skills to reach alone its object" (Leitão 2009). Since agents represent physical and virtual objects that can be networked in a distributed, modular and flexible manner, it is possible to design complex re-configurable systems. As well as the I4.0 Components industrial agents can be structured in a hierarchical manner according to the Holonic principles (Colombo, Neubert, and Schoop 2001; Leitão 2009) and representing system/subsystem relations.

However, several challenges should be addressed in order to enhance the utilization of MAS technology in the industrial automation domain properly:

- Requirements, design, deployment, and assessment of industrial agent systems.
- Integration of industrial agents within the service-oriented architectures and cloud-based industrial infrastructures.
- Focus on real-time control layer supporting industrial agent-based systems.
- Focus on industrial agent cybersecurity and safety that is internationally agreed.
- Harmonization of standardization goals and needs with respect to industrial agent technology among standards developing organizations and industrial activities at various standardization levels.

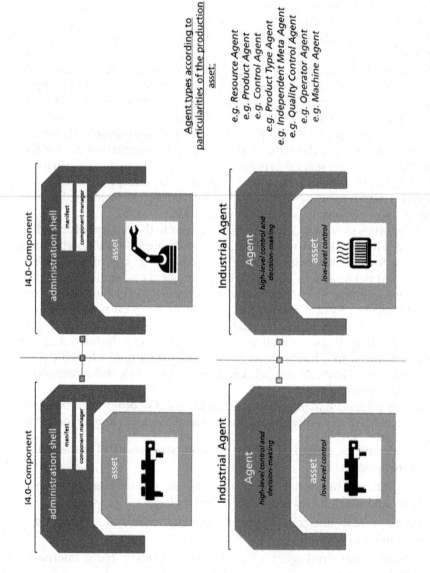

FIGURE 2.4
Industrie 4.0 Components and industrial agents. (From Fraunhofer IPA, based on Contreras, Garcia, and Diaz 2017; Leitão, Colombo, and Karnouskos 2016; VDI/VDE Society Measurement and Automatic Control 2016.)

2.2.2.4 *Connectivity*

The introduction of IoT technology (Ray 2016) into manufacturing environments opened up the landscape of connectivity solutions even further (Razzaque et al. 2016). In the past it was more common to interconnect enterprise IT systems with the help of on-premise Middleware components like Enterprise Service busses, if not hard-wired like in many cases with OT and its control and monitoring systems. A Middleware's main purpose is to integrate other software systems and components with each other. These systems often offer services for communication, integration application execution, monitoring, and operation. The main intent of Middleware is to make development of applications easier by offering common abstractions that add an abstraction layer to hide the heterogeneity of the available interfaces (Gosewehr et al. 2017; Schel et al. 2017). By adding a cloud-layer to the enterprise IT architecture and with the advent of cloud-edge architectures, new additional components have been introduced. Machines either get their own "cloud-adapters" to enable cloud connectivity. These adapters are often referred to as gateways, which very often provide interfaces to the either proprietary standard of a manufacturer or a common interface which supports a standardized open communication protocol and message format. To enable SOA applications and to bridge the communication from the closed shop floor environment to a cloud, these adapters then act as gateways to cloud-based systems, either private, hybrid, or public (Colombo et al. 2014). Cloud systems normally offer standardized and harmonized interfaces using open standards, so that connectivity can be achieved directly or by using gateway technology (Shah, Bhatt, and Patel 2018). These interfaces are often part of an integration layer, which can be based on Middleware technology. Depending on the envisioned architecture, Middleware-based integration layers can be located on-premise, in the cloud, or in a hybrid fashion. While most Middleware-, gateway-, and cloud-interfaces nowadays are built on open standard protocols (Groves, Yan, and Weiwei 2016), the main problems when integrating these systems stem from varying messages formats and semantics. Middleware components are normally capable of mapping different message formats, but this process often needs to be configured manually (Givehchi et al. 2017).

Gateways are often used as legacy integration solutions for equipment (Faul, Jazdi, and Weyrich 2016). Most equipment manufacturers nowadays offer a gateway solution to enable cloud-connectivity for their products. Newer products tend to provide native connectivity with the help of embedded computing devices, either providing a standard communications protocol solution like an OPC-UA server or the capability to build your own connectivity (Haskamp et al. 2017). Many third-party providers are offering solutions that are based on IoT technology and embedded devices like the Raspberry Pi computing module. Gateways can also be parts of a hybrid infrastructure and act as distributed parts of an integration layer solution and offer the ability to host services and computing capability for data processing. For solutions like these, best practices and reference architecture models (Weyrich and Ebert 2016) are reasonable steps to harmonize the landscape.

2.2.2.5 Interoperability

For systems interoperability next to semantics and data models the most important factors are communication protocols and connectivity standards (Groves, Yan, and Weiwei 2016). These allow the exchange of data video messages. The most important ones nowadays are IP-based (network layer) communication protocols built on top of physical communication technology like Ethernet, Wireless Local Area Network (WLAN), Personal Area Network (PAN), mobile wireless (2G, 3G/UMTS, 4G/LTE), and Wireless Wide Area Network (WWAN). Transmission Control Protocol (TCP) and User Datagram Protocol (UDP) are the standard transport layers, on top of which today's most important communication protocols are working on. The Data Distribution Service (DDS) Framework[1] is built on the DDS Interoperability Wire Protocol (DDSI-RTPS) transport protocol, oneM2M[2] works with CoAP.[3] MQTT,[4] and HTTP,[5] while OPC-UA[6] supports HTTP and its proprietary UPC-UA binary protocol. Web services are mainly built on the HTTP transport protocol often with either the older Simple Object Access Protocol (SOAP)[7] and the nowadays ubiquitous REST,[8] even enabling bidirectional communication with Websocket[9] built on top of HTTP, or the upcoming HTTP/2.[10] For interoperability of industrial applications, currently OPC-UA has found widespread acceptance and has been adapted by many manufacturers to enable connectivity and easy interoperability. But due to the current OPC-UA architecture its application is not the best fit for IoT application (e.g., only recent support for PubSub,[11] which is why other standards like oneM2M have been created to ensure the required connectivity and even semantic interoperability for future IIoT applications.

Most current technology for communication is very mature at the moment. One of the most important upcoming standards is the Time Sensitive Networks (TSN)[12] for Ethernet and 5G[13] communication, which will enable the

---

[1] http://www.omg.org/spec/DDS/
[2] http://www.onem2m.org/
[3] http://coap.technology/
[4] http://mqtt.org/
[5] https://tools.ietf.org/html/rfc2616
[6] https://opcfoundation.org/about/opc-technologies/opc-ua/
[7] https://www.w3.org/TR/soap/
[8] https://www.w3.org/2001/sw/wiki/REST
[9] https://tools.ietf.org/html/rfc6455
[10] https://tools.ietf.org/html/rfc7540
[11] https://opcfoundation.org/news/press-releases/opc-foundation-announces-opc-ua-pubsub-release-important-extension-opc-ua-communication-platform/
[12] http://www.ieee802.org/1/pages/tsn.html
[13] http://www.3gpp.org/release-15

reliability needed for time and safety critical industrial applications, which with the current standards cannot be sufficiently implemented.

2.2.2.6 Semantics and Data Models

This group describes standardization needs with the focus on semantics and information modeling.

Reconfigurable production systems and application tools are interconnected units that exchange information over the whole product life cycle. The dynamics changes in a product life cycle, especially for a family of various products, presents a complex manufacturing scenario, in which a system has been reconfigured in order to align with production demands. Within the life cycle, connected systems follow complex communication scenarios and patterns, for example, Machine-to-Machine or Machine-to-Human communication, and therefore need to exchange information that can be understood by other involved participating production units and applications.

Furthermore, industrial vocabularies and terminology are very complex subjects, usually describing data generated or consumed by machines as well as data with information about various components of these machines. Hence, to support the seamless information exchange, uniform semantics is required. This important goal can be achieved only under certain conditions that are usually limited by syntactical, semantical, and pragmatic linguistic aspects (German Standardization Roadmap 2018), for example, the existence of the common vocabulary and harmonized semantic payloads, syntactical synchronization of the conversational topics, and regulation of a language function in interaction between a human and a machine.

Data models and database schema are inherent pragmatic components of a query language. It is important that a standardized database schema and data structures that are implied by the schema can be understood and applied by system developers and system integrators.

There is already a wide range of work with regard to standardization of common ontology and other important issues concerning the information and data models (e.g., ETSI TR 101 584, ETSI TS 103 264, and ETSI TS 103 267) that could be applicable to industrial scenarios as well as relative standards of metadata models for an organization (ISO/IEC 11179), integration of life cycle data for process plants including oil and gas production facilities (ISO 15926), semantics-based requirements and framework of the IoT (Recommendation ITU-T Y.2076), etc. However, for advanced standardization needs, the topic of the universe semantics for I4.0 application appears to be a rather specialized field of study till the present days.

Within this context PERFoRM identifies the following standardization needs:

- Harmonization of heterogeneous semantics and data models with regard to I4.0.
- Focus on development of a common vocabulary.

- Review requirements focusing on semantic, syntactic, and pragmatic aspects of a language in context of I4.0.
- Introduce concepts for object-oriented database models to respond the conceptual modeling tools (e.g., Unified Modeling Language [UML]).
- Elaborate on data structures for cross-domain data exchange models.
- Joint work with the organizations developing I4.0-related industrial communication protocols with regard to apply uniform data models to achieve a high degree of flexibility in the domain.

2.2.2.7 Applications

I4.0 (RAMI4.0-compliant) solutions strongly rely on the use of industrial applications. Thus, this group focuses on software applications and production processes.

Virtualization and servitization (Baines and Lightfoot 2013; Colombo et al. 2014, 2017) are key topics in smart manufacturing domain. Services have already reached a sufficient basis of standardization relating to the service-level agreement, metrics, and information modeling structures.

However, at the international level there is a gap regarding application of services at the federal basis as well as portability of services between various manufacturing cloud ecosystems. However, in order to provide a sufficient IT security and guarantee secure data integrity in a federative structure, a well-defined software architecture is needed. One of such standardized conceptual solutions and associated is the SOA. The SOA allows composition of compact services that can be orchestrated or choreographed (see, e.g., *SOCRADES Introducing a Service Oriented Infrastructure for Industry*, EU FP6 SOCRADES Consortium 2009). Such architecture is modular and can be easily scaled by creating additional service instances (Porrmann, Essmann, and Colombo 2017). Therefore, contributions to such specific areas as service interoperability and integration of high-level production processes and tools, in particular focusing on adapter-based solutions, are required. Seamless interoperability between cyber-physical production systems and processes can greatly contribute to the development of plug-and-produce concepts.

Additionally, the following standardization activities are required within this area:

- Review and harmonization of existing standards relating to service interoperability.
- Contribution to the industrial service requirements.
- Development of standardized approaches and use cases for service orchestration, federation, and portability.
- Specification of data flows between plant systems (OT) and business applications (IT).

2.2.2.8 Human in the I4.0 Environment

This group of gaps focuses on the role of a human in I4.0 and specific functional requirements, for example, HMI requirements.

With the introduction of CPS, workers have to deal with more complex tasks and operate with more advanced hard- and software applications. Therefore, education of specific skills by a worker is required. Furthermore, the manufacturing environment must be adapted to a human-friendly environment, that is, work process, human characteristics, and connected systems must be designed jointly to reflect the needs of a worker (Colombo, Schleuter, and Kircher 2015).

In the vision of I4.0 there are no physical barriers on the shop floor between robots and humans. Thus, safety and security are becoming the key themes in this area. Therefore, related standards regarding the human-machine collaboration in this context must be reviewed. The developed assistance systems and automation solutions, helping a worker and taking over the heavy tasks, must undergo strict safety regulations before being used in the production.

Additionally, the following needs could be identified:

- Reduction of complexity regarding HMI.
- Definition of I4.0-related use cases involving human factor.
- Harmonization of requirements with focus on usability, accessibility, and data representation.

2.3 Conclusion

Influenced by the digital transformation and the rapid increase of innovative technologies, the European industry and Standards Developing Organizations (SDOs) have already recognized the necessity to revise and develop new I4.0-compliant standards. Even though a large variety of standardization activities have been launched to push the new I4.0-complient standards since the last couple of years, a significant lack of synchronization and harmonization among these can be currently observed. Therefore, in order to eliminate a vast variety of similar and inconsistent standards that have already appeared on the market, the SDOs and other industrial initiatives need to assess the current situation and start standardization discussions and cross-linking activities. Though Industry 4.0 is a complex topic, there are some basic crucial gaps that need to be addressed first as, for instance, the focus and identification of standardization objectives and fields, specification of relative standardization gaps, and harmonization and compatibility among existing and upcoming standards at various standardization levels.

In this chapter, an overview on the current technology trends and standardization activities in the context of Industry 4.0 and related to the PERFoRM approach has been presented. Furthermore, based on the practical experience of the PERFoRM project (EU HORIZON2020 FoF PERFoRM 2015–2018), a list of the current challenges and needs in the standardization landscape both at European and international levels has been summarized.

References

AIOTI, Alliance for Internet of Things Innovation. June 30, 2017. "AIOTI Strategy 2017–2021." Accessed August 09, 2018. https://aioti.eu/wp-content/uploads/2017/11/AIOTI_Stategy_2017-2021_V1.0_FINAL_WEB.pdf.

Baines, Tim, and Howard Lightfoot. 2013. *Made to Serve: How Manufacturers can Compete Through Servitization and Product-Service Systems*. Chichester, West Sussex: John Wiley & Sons Inc.

Bauer, Martin, Nicola Bui, Francois Carrez, Pierpaolo Giacomin, Stephan Haller, Edward Ho, et al. 2014. "Introduction to the Architectural Reference Model for the Internet of Things: Internet of Things—Architecture." Architectural Reference Model. https://iotforum.org/wp-content/uploads/2014/09/120613-IoT-A-ARM-Book-Introduction-v7.pdf.

Bitkom, VDMA, ZVEI. 2016. "Implementation Strategy Industrie 4.0: Report on the Results of the Industrie 4.0 Platform." https://www.zvei.org/fileadmin/user_upload/Presse_und_Medien/Publikationen/2016/januar/Implementation_Strategy_Industrie_4.0_-_Report_on_the_results_of_Industrie_4.0_Platform/Implementation-Strategy-Industrie-40-ENG.pdf.

Cala, A., M. Foehr, D. Rohrmus, N. Weinert, O. Meyer, M. Taisch, et al. 2016. "Towards industrial exploitation of innovative and harmonized production systems." Proc. of the 42nd Annual Conference of the *IEEE Industrial Electronics Society (IECON 2016)*, pp. 5735–5740.

Calà, Ambra, Arndt Luder, Ana Cachada, Flávia Pires, José Barbosa, Paulo Leitão, et al. 2017. "Migration from traditional towards cyber-physical production systems." Proc. of the 15th IEEE Int. Conf. on Industrial Informatics INDIN2017, 1147–1152.

"CEN and CENELEC's New Focus Group on Blockchain and Distributed Ledger Technologies (DLT)." 2017. News release. December 14. Accessed August 09, 2018. https://www.cencenelec.eu/news/articles/Pages/AR-2017-012.aspx.

Cisco Systems, Inc., Cisco Systems (USA) Pte. Ltd., Cisco Systems International BV Amsterdam. 2014. "The Internet of Things Reference Model: DRAFT - Control Distribution." White Paper. Accessed September 08, 2018. http://cdn.iotwf.com/resources/71/IoT_Reference_Model_White_Paper_June_4_2014.pdf.

Colombo, A. W., R. Neubert, and R. Schoop. 2001. "A Solution to Holonic Control Systems." In *IEEE International Conference on Emerging Technologies and Factory Automation 2001*. 489–498. Piscataway: IEEE.

Colombo, A. W., R. Schoop, and R. Neubert. 2006. "An agent-based intelligent control platform for industrial holonic manufacturing systems." *IEEE Trans. Ind. Electron.* 53 (1): 322–337. doi:10.1109/TIE.2005.862210.

Colombo, Armando W., Dirk Schleuter, and Matthias Kircher. 2015. "An Approach to Qualify Human Resources Supporting the Migration of SMEs into an Industrie4.0-Compliant Company Infrastructure." In *IECON 2015—Yokohama: 41st Annual Conference of the IEEE Industrial Electronics Society: November 9–12, 2015, Pacifico Yokohama, Yokohama, Japan,* edited by Kiyoshi Ohishi, and Hideki Hashimoto, 3761–3766. Piscataway, NJ: IEEE.

Colombo, Armando W., Stamatis Karnouskos, Okyay Kaynak, Yang Shi, and Shen Yin. 2017. "Industrial cyberphysical systems: A backbone of the fourth industrial revolution." *IEEE Ind. Electron. Mag.* 11 (1): 6–16. doi:10.1109/MIE.2017.2648857.

Colombo, Armando W., Stamatis Karnouskos, Yang Shi, Shen Yin, and Okyay Kaynak. 2016. "Industrial cyber–physical systems [scanning the issue]." *Proc. IEEE.* 104 (5): 899–903. doi:10.1109/JPROC.2016.2548318.

Colombo, Armando W., Thomas Bangemann, Stamatis Karnouskos, Jerker Delsing, Petr Stluka, Robert Harrison, et al. 2014. *Industrial Cloud-Based Cyber-Physical Systems.* Cham: Springer International Publishing.

Contreras, Juan David, Jose Isidro Garcia, and Juan David Diaz. 2017. "Developing of Industry 4.0 applications." *Int. J. Onl. Eng.* 13 (10): 30. doi:10.3991/ijoe.v13i10.7331.

DIN. 2018. *Deutsches Institut für Normung e. V.* Accessed September 08, 2018. https://www.din.de/en.

DIN SPEC 91345:2016-04. 2016. *Reference Architecture Model Industrie 4.0 (RAMI4.0).* https://www.beuth.de/en/technical-rule/din-spec-91345-en/250940128.

DIN. 2015. "Standardization at Various Levels." Accessed October 10, 2018. https://www.din.de/en/din-and-our-partners/din-in-europe/european-standardization.

DIN/VDE. 2017. "DIN EN 62890:2017-04." Accessed October 10, 2018. https://www.beuth.de/de/norm-entwurf/din-en-62890/269121145.

DKE/VDE. 2018. *DKE Deutsche Kommission Elektrotechnik Elektronik Informationstechnik in DIN und VDE.* Accessed September 08, 2018. https://www.dke.de/de.

EU FP6 SOCRADES Consortium. 2009. *SOCRADES Introducing a Service Oriented Infrastructure for Industry.* Accessed October 10, 2018. https://www.youtube.com/watch?v=BCcqb8cumDg, https://www.youtube.com/watch?v=K8OtFD6RLMM or https://www.youtube.com/watch?v=K8OtFD6RLMM&t=404s.

"EU HORIZON2020 FoF PERFoRM 2015-2018." Accessed October 10, 2018. https://cordis.europa.eu/project/rcn/198360_de.html.

EU MANUFUTURE. 2013. Accessed October 10, 2018. http://www.manufuture2013.eu/images/MANUFUTURE2013_Catalogue.pdf.

Faul, Alexander, Nasser Jazdi, and Michael Weyrich. 2016. "Approach to Interconnect Existing Industrial Automation Systems with the Industrial Internet." In *2016 IEEE 21st International Conference on Emerging Technologies and Factory Automation (ETFA): September 6–9, 2016 Berlin, Germany,* 1–4. Piscataway, NJ: IEEE.

François Coallier. 2018. "A Strategic View of ISO/IEC JTC 1/SC41 IoT and Related Technologies." Accessed October 10, 2018. https://www.itu.int/en/ITU-T/Workshops-and-Seminars/20180604/Documents/Francois_Coallier_P_V2.pdf.

"German Standardization Roadmap: Industrie 4.0." 2018. DIN/DKE—Roadmap. Accessed August 03, 2018. https://www.din.de/blob/65354/57218767bd6da1927b181b9f2a0d5b39/roadmap-i4-0-e-data.pdf.

"German Standardization Roadmap: Industry 4.0." 2016 Version 2 https://www.din.de/blob/65354/f5252239daa596d8c4d1f24b40e4486d/roadmap-i4-0-e-data.pdf. Accessed February 22, 2018.

Givehchi, Omid, Klaus Landsdorf, Pieter Simoens, and Armando W. Colombo. 2017. "Interoperability for industrial cyber-physical systems: An approach for legacy systems." *IEEE Trans. Ind. Inf.* 13 (6): 3370–3378. doi:10.1109/TII.2017.2740434.

Gosewehr, Frederik, Jeffrey Wermann, Waldemar Borsych, and Armando Walter Colombo. 2017. "Specification and design of an industrial manufacturing middleware." Proc. of the 15th IEEE Int. Conf. on Industrial Informatics INDIN2017, pp. 1160–1166.

Groves, Christian, Lui Yan, and Yang Weiwei. 2016. "Overview of IoT Semantics Landscape." Accessed August 09, 2018. https://www.iab.org/wp-content/IAB-uploads/2016/03/IoTSemanticLandscape_HW_v2.pdf.

Haskamp, Hermann, Michael Meyer, Romina Mollmann, Florian Orth, and Armando W. Colombo. 2017. "Benchmarking of existing OPC UA implementations for Industrie 4.0-Compliant digitalization solutions." Proc. of the 15th IEEE Int. Conf. on Industrial Informatics INDIN2017, pp. 589–594.

Herrera, V. V., A. Bepperling, A. Lobov, H. Smit, A. W. Colombo, and J. L. M. Lastra. 2008. "Integration of Multi-Agent Systems and Service-Oriented Architecture for Industrial Automation." In *6th IEEE International Conference on Industrial Informatics, 2008: INDIN 2008; Daejeon, South Korea, 13–16 July 2008, 768–773.* Piscataway, NJ: IEEE Service Center.

IEC 61512. 1995–2006. "IEC 61512, ANSI/ISA-88." Accessed October 10, 2018. http://deacademic.com/dic.nsf/dewiki/642022.

IEC PAS 63088:2017. 2017. *Smart manufacturing - Reference Architecture Model Industry 4.0 (RAMI4.0).* 1.0th ed. 25.040.01 - Industrial Automation Systems in General 35.080 - Software 35.240.50 - IT Applications in Industry. https://webstore.iec.ch/publication/30082.

IEEE. 2018. *About IEEE.* Accessed September 08, 2018. https://www.ieee.org/about/index.html.

"IEEE and National Institute of Standards and Technology (NIST) Team on Standards Development for Intercloud Interoperability and Federation: Collaboration between NIST and IEEE P2302™ Will Help Build Consensus on Creating an Intercloud—An Open, Transparent Infrastructure amongst Cloud Providers to Support Evolving Technological and Business Models." 2017. News release. July 25. Accessed August 09, 2018. http://standards.ieee.org/news/2017/intercloud_interoperability_and_federation.html.

IEEE-SA. "2660.1 - Recommended Practices on Industrial Agents: Integration of Software Agents and Low Level Automation Functions." News release. Accessed September 08, 2018. https://standards.ieee.org/develop/project/2660.1.html.

IERC, European Research Cluster on the Internet of Things. January. "Internet of Things: Position Paper on Standardization for IoT technologies." European Reserach Cluster on the Internet of Things. Accessed August 09, 2018. http://www.internet-of-things-research.eu/pdf/IERC_Position_Paper_IoT_Standardization_Final.pdf.

IIC. 2017. "The Industrial Internet of Things: Volume G1: Reference Architecture." IIC:PUB:G1:V1.80:20170131. Accessed August 09, 2018. https://www.iiconsortium.org/IIC_PUB_G1_V1.80_2017-01-31.pdf.

IIC. 2018. "Industrial Internet Consortium." Accessed September 08, 2018. https://www.iiconsortium.org/.

IIC-IVI. 2017. "The Industrial Internet Consortium and the Industrial Value Chain Initiative Sign MoU." 2017. News release. April 26. Accessed August 09, 2018. https://www.iiconsortium.org/press-room/04-26-17.htm.

IIoT. 2018. "The Industrial Internet of Things." Accessed October 10, 2018. https://www.iiconsortium.org/IIC_PUB_G1_V1.80_2017-01-31.pdf.

International Organization for Standardization. 2016. "Information Technology—Internet of Things Reference Architecture (IoT RA)." Accessed October 10, 2018. http://ivezic.com/iot-security-standards-frameworks-guidelines/iso-internet-of-things-reference-architecture/.

ISO/IEC. ISO/IEC CD 30141: Information technology - Internet of Things Reference Architecture. Accessed September 08, 2016. https://www.iso.org/standard/65695.html.

ITU-T, Telecommunication Standardization Sector of ITU. 2013. "Overview of the Internet of Things: Recommendation ITU-T Y.2060." Next Generation Networks—Frameworks and Functional Architecture Models. Accessed August 09, 2018. https://www.itu.int/en/ITU-T/gsi/iot/Pages/default.aspx.

IVI 2018-1. *Industrial Value Chain Initiative*. Accessed September 08, 2018. https://iv-i.org/wp/en/.

IVI 2018-2, Industrial Value Chain Initiative. 2018. "IVRA Next: Strategic Implementation Framework of Industrial Value Chain for Connected Industries." Industrial Value Chain Reference Architecture—Next. Accessed September 08, 2018. https://iv-i.org/wp/wp-content/uploads/2018/04/IVRA-Next_en.pdf.

Kagermann, H., W. Wahlster, and J. Helbig. 2013. "Recommendations for Implementing the Strategic Initiative INDUSTRIE 4.0: Final Report of the Industrie 4.0 Working Group." Securing the Future of German Manufacturing Industry. Accessed April 25, 2018.

Koren, Y., U. Heisel, F. Jovane, T. Moriwaki, G. Pritschow, G. Ulsoy, and H. van Brussel. 2013. "Reconfigurable Manufacturing Systems." In *Manufacturing Technologies for Machines of the Future: 21st Century Technologies*, edited by Anatoli I., Daščenko. Softcover reprint of the hardcover 1. ed. 2003, 627–665. Berlin: Springer.

Kortuem, Gerd, Fahim Kawsar, Vasughi Sundramoorthy, and Daniel Fitton. 2010. "Smart objects as building blocks for the internet of things." *IEEE Internet Comput.* 14 (1): 44–51. doi:10.1109/MIC.2009.143.

Leitão, Paulo. 2009. "Agent-based distributed manufacturing control: A state-of-the-art survey." *Eng. Appl. Artif. Intel.* 22 (7): 979–991. doi:10.1016/j.engappai.2008.09.005.

Leitão, Paulo, Armando W. Colombo, and Francisco Restivo. 2006. "A formal specification approach for holonic control systems: The ADACOR case." *IJMTM* 8 (1/2/3): 37. doi:10.1504/IJMTM.2006.008790.

Leitão, Paulo, Armando Walter Colombo, and Stamatis Karnouskos. 2016. "Industrial automation based on cyber-physical systems technologies: Prototype implementations and challenges." *Comput Ind.* 81: 11–25. doi:10.1016/j.compind.2015.08.004.

Leitão, Paulo, José Barbosa, Arnaldo Pereira, José Barata, and Armando W. Colombo. 2016. "Specification of the PERFoRM architecture for the seamless production system reconfiguration." Proc. of the 42nd Annual Conference of the IEEE Industrial Electronics Society (IECON 2016), pp. 5729–5734.

Li, Qing, Qianlin Tang, Iotong Chan, Hailong Wei, Yudi Pu, Hongzhen Jiang, Jun Li, and Jian Zhou. 2018. "Smart manufacturing standardization: Architectures, reference models and standards framework." *Comput. Ind.* 101: 91–106. doi:10.1016/j.compind.2018.06.005.

Lin, Shi-Wan, Brett Murphy, Erich Clauer, Ulrich Loewen, Ralf Neubert, Gerd Bachmann, Madhusudan Pai, and Martin Hankel. 2017. "Architecture Alignment and Interoperability: An Industrial Internet Consortium and Plattform Industrie 4.0 Joint Whitepaper." IIC:WHT:IN3:V1.0:PB:20171205. Accessed August 09, 2018. https://www.iiconsortium.org/pdf/JTG2_Whitepaper_final_20171205.pdf.

LNI 4.0. 2018. *Labs Network Industrie 4.0—Praxiserprobungen und Standards.* Accessed September 08, 2018. https://lni40.de/.

Lu, Yan, K. C. Morris, and Simon Frechette. 2016. *Current Standards Landscape for Smart Manufacturing Systems.* National Institute of Standards and Technology. Accessed February 22, 2018.

Lu, Yan, KC Morris, and Simon Frechette. 2016. "Current Standards Landscape for Smart Manufacturing Systems." NISTIR 8107. Accessed August 09, 2018. https://nvlpubs.nist.gov/nistpubs/ir/2016/NIST.IR.8107.pdf.

McKinsey & Company. 2015. "Industry 4.0 How to Navigate Digitization of the Manufacturing Sector." Accessed 10, 2018. http://worldmobilityleadershipforum.com/wp-content/uploads/2016/06/Industry-4.0-McKinsey-report.pdf.

Meyer, Olga, Greg Rauhoeft, Christian Henkel, and Ambra Calà. 2017. "A harmonized approach for constructing a robust and efficient technology backbone for agile manufacturing systems." Proc. of the 15th IEEE Int. Conf. on Industrial Informatics INDIN2017, pp. 1111–1116.

Monostori, László. 2014. "Cyber-physical production systems: Roots, expectations and R&D challenges." *Procedia CIRP* 17:9–13. doi:10.1016/j.procir.2014.03.115.

Nagorny, Kevin, Armando W. Colombo, and Uwe Schmidtmann. 2012. "A service- and multi-agent-oriented manufacturing automation architecture." *Comput. Ind.* 63 (8): 813–823. doi:10.1016/j.compind.2012.08.003.

NAMUR. Accessed September 08, 2018. https://www.namur.net/.

NIST. 2018. *National Institute of Standards and Technology.* Accessed September 08, 2018 https://www.nist.gov/.

OCF. 2018. *Open Connectivity Foundation.* Accessed September 08, 2018. https://open-connectivity.org/.

oneM2M. *Standards for M2M and the Internet of Things.* 08-February-2018. Joint IIC-oneM2M workshop: Delivering Value with IIoT: Delivering Value across Machine to Machine & Industrial IoT Sectors. Accessed August 09, 2018. http://www.onem2m.org/15-events.

oneM2M. *Standards for M2M and the Internet of Things.* Accessed September 08, 2018. http://www.onem2m.org/about-onem2m/why-onem2m.

Platform i40, 2018. 2018. "Industrie4.0." Accessed October 12, 2018. https://www.plattform-i40.de/I40/Navigation/DE/Home/home.html.

Plattform Industrie 4.0, Robot Revolution Initiative, Standardization Council Industrie 4.0. 2018. "The Common Strategy on International Standardization in Field of the Internet of Things/Industrie 4.0." https://www.plattform-i40.de/I40/Redaktion/DE/Downloads/Publikation/common-strategy-international-standardization.html.

Plattform Industrie 4.0, Standardization Council Industrie 4.0, Alliance Industrie du Futur, Ministero dello Sviluppo Economico, UNI -Ente Nazionale Italiano di Unificazione. 2018. "Paris Declaration for Smart Manufacturing—by the Working Group "Standardization and Reference Architecture" of the Trilateral Cooperation: Digitizing European Industry | Stakeholder Forum 2018 in Paris." Accessed August 09, 2018. https://www.plattform-i40.de/I40/Redaktion/DE/Downloads/Publikation/wg3-trilaterale-coop.pdf?__blob=publicationFile&v=6.

Porrmann, Thomas, Roland Essmann, and Armando W. Colombo. 2017. "Development of an Event-Oriented, Cloud-Based SCADA System Using a Microservice Architecture Under the RAMI4.0 Specification: Lessons Learned." In *Proceedings IECON 2017—43rd Annual Conference of the IEEE Industrial Electronics Society: China National Convention Center, Bejing, China, 29 October—01 November, 2017,* 3441–3448. Piscataway, NJ: IEEE.

Proceedings of the IECON2016—42nd Annual Conference of the IEEE Industrial Electronics Society. 2016. Piscataway, NJ: IEEE.

Ray, Partha Pratim. 2016. "A survey of IoT cloud platforms." *Future Computing and Informatics Journal.* 1 (1–2): 35–46. doi:10.1016/j.fcij.2017.02.001.

Razzaque, Mohammad Abdur, Marija Milojevic-Jevric, Andrei Palade, and Siobhan Clarke. 2016. "Middleware for internet of things: A survey." *IEEE Internet Things J.* 3 (1): 70–95. doi:10.1109/JIOT.2015.2498900.

Ribeiro, Luis, José Barata, and Armando Colombo. 2008. "MAS and SOA: A Case Study Exploring Principles and Technologies to Support Self-Properties in Assembly Systems." In *2008 Second IEEE International Conference on Self-Adaptive and Self-Organizing Systems Workshops.* 192–197: IEEE.

Schel, Daniel, Christian Henkel, Daniel Stock, Olga Meyer, Greg Rauhoeft, Peter Einberger, Matthias Stöhr, Marc Daxer, and Joachim Seidelmann. 2017. "Manufacturing Service Bus: An Implementation % This file Was Created with Citavi 5.7.0.0."

Schoop, R., A. W. Colombo, B. Suessmann, and R. Neubert. 2002. "Industrial Experiences, Trends and Future Requirements on Agent-Based Intelligent Automation." In *IECON-2002: Proceedings of the 2002 28th Annual Conference of the IEEE Industrial Electronics Society: Sevilla, Spain. November 5–8, 2002,* 2978–2983. Piscataway, NJ: IEEE.

SCI 4.0. 2018. *Home—Standardisation Council Industrie 4.0.* Accessed September 08, 2018. https://sci40.com/de/.

Shah, Nirali, Chintan Bhatt, and Divyesh Patel. 2018. "IoT Gateway for Smart Devices." In *Internet of Things and Big Data Analytics toward Next-Generation Intelligence.* Vol. 30, edited by Nilanjan Dey, Aboul E. Hassanien, Chintan Bhatt, Amira Ashour, and Suresh C. Satapathy, 179–198. Studies in Big Data 30. Cham: Springer International Publishing.

"Through IoT, Japanese Factories Connected Together: The "Next Innovation" in Manufacturing (monozdukuri) Is Coming!". 2015. News release. May 15. Accessed August 09, 2018. http://www.meti.go.jp/english/publications/pdf/journal2015_05a.pdf.

VDI/VDE Society Measurement and Automatic Control. 2016. "Status Report: Industrie 4.0 Service Architecture—Basic Concepts for Interability." Accessed August 22, 2018. https://www.vdi.de/fileadmin/vdi_de/redakteur_dateien/gma_dateien/END_3_Status_Report_Industrie_4_0_-Service_Architecture.pdf.

Wan, Jiafu, Hu Cai, and Keliang Zhou. 2015. "Industrie 4.0: Enabling Technologies." In *Proceedings of 2015 International Conference on Intelligent Computing and Internet of Things*. 135–140. Piscataway, NJ: IEEE.

Weyrich, Michael and Christof Ebert. 2016. "Reference architectures for the internet of things." *IEEE Software*. 33 (1): 112–116. doi:10.1109/MS.2016.20.

Wübbeke, Jost, Mirjam Meissner, Max J. Zenglein, Jaqueline Ives, and Björn Conrad. December 2016. "MADE IN CHINA 2025: The Making of a High-Tech Superpower and Consequences for Industrial Countries." Accessed August 09, 2018. https://www.merics.org/sites/default/files/2017-09/MPOC_No.2_MadeinChina2025.pdf.

Xu, Li Da, Wu He, and Shancang Li. 2014. "Internet of things in industries: A survey." *IEEE Trans. Ind. Informat.* 10 (4): 2233–2243. doi:10.1109/TII.2014.2300753.

3

PERFoRM System Architecture

Paulo Leitão, José Barbosa
(Instituto Politécnico de Bragança)

Jeffrey Wermann, Armando W. Colombo
(Institute for Informatics, Automation and Robotics (I²AR))

CONTENTS

3.1 Introduction

3.1.1 What Is an Architecture?

The definition of architecture may vary from person to person and from field to field in the sense that anyone of anything might have different perspectives of "what an architecture is?" Although in the present work one did not consider the use of ISO/IEC/IEEE 42010 (2011), it can still be used to better describe the notion of "architecture."

In such way, an architecture is defined as "… (system) fundamental concepts or properties of a system in its environment embodied in its elements, relationships, and in the principles of its design and evolution...." Better clarifying the aforementioned sentence, it can be stated that in the present case the *system* (in brackets) signifies a system that is "man-made" accordingly with the term defined in the ISO 15288, mainly focusing a broad field of senses, namely software, data, human, and processes. At its present state, the Production harmonizEd Reconfiguration of Flexible Robots and Machinery (PERFoRM) architecture intends to provide a conceptual architecture (providing its fundamentals) that, by following the principles inherited in Industry 4.0, could be used by practitioners or developers that are aiming to develop tools and/or solutions to be deployed in such environment.

Therefore, an architecture is an abstract system representation rather than an artefact. In the definition provided in the International Organization for Standardization (ISO)/International Electrotechnical Commission (IEC)/ Institute of Electrical and Electronics Engineers (IEEE) 42010, there are two possible paths in the architecture design, namely *"concepts or properties,"* whereas in the present case, the concept, being a concept in the project mind, is the best definition.

Another keyword that should be denoted in the former definition is "fundamental." In the present considerations this means that the PERFoRM architecture will define its main elements, the basic relationships among them as also the core design principles.

Matching such an approach with the classical ISA-95 layered approach, the PERFoRM system architecture will cover the different layers of the automation pyramid, and particularly the layers L2 and L3, by providing an infrastructure and methodology to deploy the new generation of automation systems in the form of distributed cloud-based automation systems, following the principles sustained by Industry 4.0 and the factories of the future. In fact, the PERFoRM system architecture should provide the means for both the vertical and horizontal system integration.

3.1.2 The PERFoRM System Architecture

The design of the system architecture for new innovative production systems should take into consideration the requirements and functionalities of the new generation of Industrial Cyber-Physical-Systems (ICPS) and more specifically Cyber-Physical-Production-Systems (CPPSs), and particularly those established for the PERFoRM use cases. An important assumption is to reuse the results from the previous successful R&D projects in the field of distributed automation control systems, instead of developing a new architecture from scratch. This assumption also guarantees that the proposed architecture will be backward compatible and aligned with the current state-of-the-art approaches, increasing therefore its industrial adoption possibility. Furthermore, it is also crucial that the proposed architecture be aligned with the current trends, particularly with those being followed by the several international initiatives, for example, the Industry4.0 platform and its Reference Architectural Model (RAMI) (ZVEI 2016).

The analysis of the identified requirements shows that the system architecture should

- be based on smart and heterogeneous production components;
- be able to support the seamless system reconfiguration;
- be able to enhance planning, simulation, and operational features; and
- be able to provide human operators with enhanced information and assistance.

As a result of these assumptions, the PERFoRM system architecture for the seamless production system reconfiguration demands the use of distributed control approaches instead of using traditional centralized ones, which are characterized to be rigid and monolithic structures that are not anymore able to face the levels of responsiveness and reconfigurability imposed by the factories of the future. For this purpose, the system architecture, as illustrated in Figure 3.1, is based on a community of smart and heterogeneous components, comprising hardware devices and software applications, addressing different ISA-95 levels (Leitão et al. 2016).

The plethora of distributed and heterogeneous hardware devices and software applications expose their functionalities as services, following service-oriented design principles, which are offered to the other components. The use of standard interfaces (in terms of syntax and semantics) and industrially adopted M2M protocols, covering the several ISA-95 automation levels, allows the interconnection of production components in a transparent manner. In addition, a distributed service-based integration layer (known as industrial Middleware) is used to ensure a transparent, secure, and reliable interconnection, addressing the interoperability in such heterogeneous ecosystems. An important innovation of this integration layer is its distributed approach, instead of the traditional centralized ones that can act as a single point of failure as well as a limitation for the system scalability.

Some production components can be enriched with intelligence and higher processing capabilities to run more complex algorithms allowing them to process higher amount of data, producing a valuable analysis to be used when needed (i.e., in a real-time basis), and also supporting the seamless reconfiguration of the system and the achievement of self-properties, for example, self-adaptation, self-diagnosis. For this purpose, artificial intelligence and machine learning (ML) techniques are embedded in these production systems, particularly data mining and Multi-agent Systems (MAS) technology (Ferber 1999; Colombo et al. 2004; Wooldridge 2009; Leitão. 2009). Advanced data analytics, for example considering data mining algorithms, can be easily integrated in production components, providing a smart layer, close to the machines, that allows the detection of possible problematic situations beforehand and the fast response to condition changes in a timely manner, improving their behavior during runtime.

Aiming to reach long-term planning and optimization, these established smart production components should be interconnected with higher ISA-95 levels. PERFoRM architecture empowers this vertical data interchange

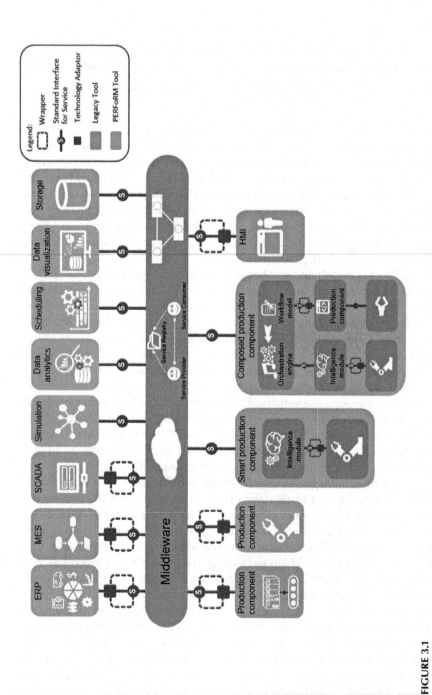

FIGURE 3.1
Overall Production harmonizEd Reconfiguration of Flexible Robots and Machinery (PERFoRM) system architecture. (From Angione et al. 2017)

by enabling a transparent data flow independently of where the data is located, promoting the use of industrial standardized protocols, for example, Open Platform Communications Unified Architecture (OPC-UA) (Haskamp et al. 2017), to seamlessly interconnect the devices. Since, usually, the older machines are not equipped with the appropriate computational power, this local processing module assumes a crucial aspect not only to run real-time monitoring and diagnostic algorithms, but also to filter the collected data that is sent over the network into a cloud environment for processing, avoiding the transmission of unnecessary raw data over the network.

From a system-of-systems perspective, some production components can offer composed services based on the aggregation and orchestration of existing atomic services provided by individual smart production components. In fact, the service composition (Peltz 2003) is the combination of atomic services, and the interaction patterns between them, to create a composed service that is offered to the other entities. In this process, the service orchestration (Jammes et al. 2005) is crucial to sequence and synchronize the execution of

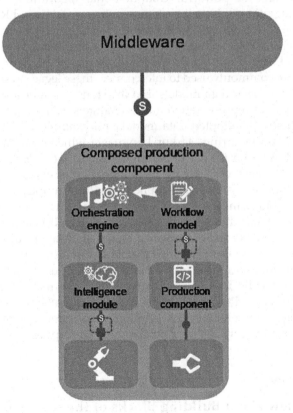

FIGURE 3.2
Orchestration of services to support the creation of Composed production components. (From EU HORIZON 2020 FoF PERFoRM 2016.)

the atomic services according to a workflow that represents the business process, providing a high-level interface for such composed process. For example, as illustrated in Figure 3.2, a new service "pick-and-place" is composed by considering the atomic service "transfer" offered by the robot device and the atomic services "open" and "close" offered by the gripper device. For this purpose, the embedded orchestration engines should synchronize the execution of the atomic services according to a process workflow, being possible to be implemented by using different formalisms (Colombo and Karnouskos 2009).

During the SOCRADES project a kind of Petri nets engine (Colombo, Carelli, and Kuchen 1997; Feldmann and Colombo 1998; Murata 1989), tailored for formalizing orchestration mechanisms, was developed and prototyped, embedding the engine into smart embedded input/output (I/O) devices of the Service-Oriented Architecture (SOA)-technology provider Schneider Electric Automation (Mendes et al. 2009).

Existing legacy systems focusing on planning, simulation, and operational features must be integrated and also coexist with advanced tools taking advantage of powerful computational algorithms and technologies. The integration of legacy systems, such as databases, ERP (Enterprise Resource Planning), MES (Manufacturing Execution Systems), and SCADA (Supervisory Control and Data Acquisition), is simplified by using standard interfaces and industrial middlewares addressing the higher ISA-95 levels.

Adapters are commonly used to interconnect these legacy systems by transforming their internal data models into the standard interfaces data model. The seamless data representation and exchange schema is reached by considering industrially adopted data models, for example, IEC 62264 B2MML (Business to Manufacturing Markup Language), which is an XML implementation of the ISA-95, for the backbone environment, and OPC-UA or IEC 62714 AutomationML data models for the machinery environment. Advanced planning, scheduling, and simulation applications, for example, developed using the MAS technology, may also be integrated, but in this case without the need to use adapters since they already follow the PERFoRM standard interfaces. These applications, as well as advanced simulation applications, could run in cloud platforms to take advantage of ubiquity and computational power.

The human integration assumes an important issue as flexibility driver, which is fostered by interfacing the humans with user-friendly human-machine interface (HMI), mobile devices (e.g., tablets and smartphones), and applying augmented reality technology, in order to provide relevant, up-to-date information and support the most appropriate intervention.

3.2 Overview about Building Blocks of the Architecture

This section overviews the core architectural building blocks allowing the fulfillment of the aforementioned requirements and functionalities.

3.2.1 Industrial Middleware

A key role in this system architecture is performed by the industrial Middleware (Gosewehr et al. 2017), which is a distributed service-based integration layer that aims to ensure a transparent, secure, and reliable interconnection of the diverse heterogeneous hardware devices (e.g., robotic cells and Programmable Logic Controllers [PLCs]) and software applications (e.g., MES and SCADA) presented at the PERFoRM ecosystem.

The common definition of a Middleware is a system or software component, which is used to connect different applications or systems. The target is to be able to establish a communication between these applications without them having to know about each other's inner structure. Middleware systems are getting increasingly important in the industrial world, where a huge amount of systems with different tasks are involved and have to work together to keep the production running. This reaches from low-level sensors, actuators, and controllers to the management systems for ERP.

An important innovation of this integration layer is its distributed and cloud approach, instead of the centralized ones that can be mostly found nowadays and can act as a single point of failure as well as a limitation for the system scalability. Additionally, this distributed integration layer handles the interconnection of these heterogeneous production components by following the service-orientation principles, where the connected heterogeneous software and hardware components are exposing their functionalities as services, which will be discovered and requested by the other components.

The Middleware will provide a set of basic functionalities, namely the service registry, the so-called Yellow Page system, which is storing the different services available in the whole system, and the service discovery that allows to discover new services. Additionally, it will take care of linking components that need to communicate and, if necessary, routing and translating the data in a transparent manner. These basic functionalities will be the core of the Middleware, as it is planned to be a more lightweight and not too complex system. Any kind of sophisticated intelligence, such as orchestration engines for the services, marshal use of redundant services, and services life cycle monitoring, are not included in the basic concept of the Middleware. Instead the Middleware is designed to have open interfaces to add more functionalities and intelligence such as Plug-Ins or Add-Ons.

3.2.2 Standard Interfaces

Aligned with the general vision for the Industry 4.0 platform, one of the key challenges that the PERFoRM architecture aims to tackle is the interoperability in real industrial environments, dealing with the representation and seamless exchange of data originating from a wide array of entities, often from different, albeit related, actions levels. The interconnection of heterogeneous legacy hardware devices (e.g., robots and the respective controllers) and software applications (e.g., databases, SCADA applications, and other management, analytics and logistics tools) is one of the main goals currently being pursued in this vision.

To this effect, the PERFoRM architecture employs the adoption of standard interfaces as the main drivers for pluggability and interoperability, aiming at enabling the connection between such devices and applications in a seamless and transparent manner. These interfaces should support the devices, tools, and applications with the means to fully expose and describe their services in a unique, standardized, and transparent way to enhance the seamless interoperability and pluggability, fully specifying the semantics and data flow involved in terms of inputs and outputs required to interact with these elements. Therefore, from the system point of view, both the standard interface specification and its development abstractthe underlying function operation, making transparent the way how the several architectural modules interact and operate.

These interfaces should provide a set of functionalities related to a standardized service invocation (note that an important requirement for the design of standard interfaces is the usage of service orientation to expose the device/system functionalities as services):

- The definition of the list of services to be implemented by the interface.
- The contract implementation of each service (i.e., the name, input parameters, and output parameters).
- The definition of the data model handled by the services.

For this purpose, a common data model is also adopted, serving as the data exchange format shared between the PERFoRM-compliant architectural elements. This data model covers the semantic needs associated to each entity, which in the particular case of industrial automation, means that the requirements related to each ISA-95 layer and their respective needs are considered. In this context, two particular data abstraction levels are taken into account, more specifically the machinery level, covering mainly layers L1 (automation control) and L2 (supervisory control), and the data backbone level, which covers layers L3 (manufacturing operations management) and L4 (business planning and logistics).

As such, full interoperability and harmonization of data at a system-of-systems level is achieved by coupling the standard interfaces with the data model for a common representation of data and system semantics. However, taking into account the integration of legacy devices and their own individual data models and semantic requirements, the addition of technology adapters is also required in order to enable the translation and mapping of legacy data into the common PERFoRM representation, allowing for these devices to be conferred additional intelligence and integrated into the Industrial Cyber-Physical-System paradigm.

3.2.3 Technological Adapters

As previously referred, manufacturing companies are usually characterized by the use of legacy and heterogeneous systems for the management and the execution of their production process; for example, robots, PLCs, databases, and MES systems. The innovative PERFoRM architecture can only be industrially accepted and really adopted if the possibility to integrate the legacy systems is presented. For this reason, technology adapters are key elements to connect legacy systems to the PERFoRM Middleware and to transform the legacy data model into the standard interface data model, that is, masking the legacy systems' data/functionalities according to the PERFoRM standard interfaces. In this way, the technological adapters are only necessary when there is the need to connect a legacy component (e.g., an existing database or robot) to the PERFoRM system.

Although at its essence the adapter function is to convert legacy data (non-PERFoRM compliant) into the PERFoRM standardized data model, technological adapters can be categorized into three categories, according to its scope, namely to interconnect standard legacy systems, real-time legacy systems, and HMI legacy systems. These adapters respond to the different types of legacy systems that can be found in a production environment and are able to seamlessly connect these systems with the industrial Middleware and the higher level of the enterprise network. Real-Time constraints are particularly important when considering shop floor machinery as they may need quick adjustments and corrections according to the data acquired from low-level sensors locally installed in the production resource (e.g., vibration analysis of spinning spindles). In the same manner, the interconnection with existing HMIs is crucial, not only for monitoring and controlling the production resource but also for capturing human expert knowledge and support following human activities from past experience (e.g., changeover and ramp-up operations can be supported by policies derived from past cases).

3.2.4 Human Integration

The integration of the human in the loop is seen as a key factor to improve flexibility. The analysis of human integration requirements results in a set of some recommendations with implications for the planning and designing the human-machine and human-human interfaces, addressing two different levels: At strategic level, for example, supporting decision-makers to take strategic decisions (or also considered as the "human in the mesh"), and also at operational level, for example, supporting operators or maintenance engineers to perform their tasks (or also known as "human in the loop") (EU HORIZON 2020 FoF PERFoRM 2015, 2017). In particular, the following functionalities are considered and supported in the PERFoRM architecture:

- Consultation among team members and/or with other experts, supporting virtual presence and/or the possibility to share the view, the screen, the information, voice, and chatting space. These interfaces would allow to show a detail of a part, machine, etc. to share the information displayed on a screen, to attract the attention on a particular sound, so that other colleagues can be consulted, also if they are not physically present on the shop floor.

- Acquisition of commands and data concerning human task execution, for example, gesture recognition, and provision of feedback and alerts in case of errors.
- Delivery of condition-based instructions/on-the-job training, using multimodal interfaces, offered in mobility and, in some cases, providing augmented reality functionalities.
- Attracting the attention of and alerting the operators in case of unexpected/anomalous events or behaviors and problems in the manufacturing processes and systems.
- Providing mobile turn-by-turn guidance to navigate the factory to retrieve tools, equipment, and spare parts.
- Supporting visual inspection by providing information gathered by sensors and other edge-devices.
- Displaying the dynamics and results of simulations and rescheduling, allowing the intuitive representation of alternatives and trade-offs, the comparison and evaluation of multiple objectives and performances, and facilitating collaboration and negotiation within and among teams.

In this architecture, the hardware devices (e.g., tablets, smart phones, glasses, or gesture devices), providing the described HMI functionalities (e.g., collecting data from the user or providing data to the user), are interconnected to the PERFoRM ecosystem through the industrial Middleware by using proper adapters and standard interfaces.

3.2.5 Advanced Tools for Planning, Scheduling, and Simulation

Tools particularly designed with advanced algorithms and technologies to support the production planning, scheduling, and simulation may improve the system performance and reconfigurability. These tools should be PERFoRM compliant, that is, following the service orientation and using the PERFoRM native interfaces.

These advanced tools can be clustered in three different groups: simulation solutions, planning logic solutions, and intelligent decision support and visualization solutions.

Within the simulation cluster, different technologies can be used, namely the Simulation Environment (SE) covering different interfaces to ensure the seamless integration of specific simulation models to dynamically acquire and provide data. Four different interfaces were identified:

- Getting data from the Middleware into the SE.
- Sending data from the SE to the Middleware.
- Getting data from control planning logic into the SE.
- Internal interface within the SE for the execution of the simulation model itself (actual simulation tools are Plant Simulation and AnyLogic).

Furthermore, actual simulation tools can be enabled to model flexible and reconfigurable production systems, for example, with agent-based approaches within a classical discrete event simulation tool. As a last technology, validation methods can be developed to ensure to ensure that the connection between the physical and the cyber levels harmonizes seamlessly.

The planning logic cluster includes solutions regarding the dynamic production (re-)scheduling, planning, and reconfiguration mechanisms, supporting flexible and reconfigurable manufacturing systems; for example, used in the production planning for including reconfigurable robot cells. Furthermore, these solutions include approaches for multi-objective planning and agent-based approaches. It will be possible for the results of these tools, for example, potential operation schedules and reconfiguration proposals, to be evaluated using the SE. In such a way, these tools can be harmonized with the production system and used to guide its optimal operation.

The intelligent decision support and visualization cluster is concerned with data extraction and deriving predictions. This includes predictive maintenance tools, using, for example, regression analysis, Bayesian networks, clustering, time-frequency analysis, and other methods. Moreover, what-if games, the link to the planning logic, and simulation cluster provide decision support on the manufacturing system level. In the visualization subcluster, advanced tools can enable the visualization of different possibilities of visualizations utilizing data available from the Middleware as well as from other advanced tools. Possible visualizations include value stream maps, key performance indicators (KPIs), work orders, worker-specific data, topologies, and possible what-if scenarios.

In the PERFoRM architecture, these tools are interconnected to the system by means of the Middleware and are able to connect to the data available in the system, either for gathering data for the simulation or by generating new data to be used elsewhere by other tools. Naturally, these tools will use the PERFoRM native language and consequently they will not need to use the technological adapters. Nevertheless, legacy modeling and simulation tools are pluggable by means of the use of proper adapter and standard interface.

Two exemplary process flows are provided to illustrate the data flow when interconnecting these kinds of tools (note that the actual process flow of similar tools is dependent on the tool implementation strategy itself, since the

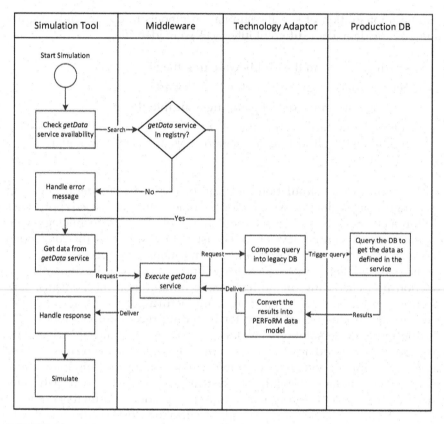

FIGURE 3.3
Exemplary flow diagram for a simulation tool integrated in the Production harmonizEd
Reconfiguration of Flexible Robots and Machinery (PERFoRM) ecosystem. (From EU
HORIZON 2020 FoF PERFoRM 2016.)

PERFoRM architecture does not bind to a particular process flow strategy).
Figure 3.3 illustrates the interconnection of a simulation tool to the PERFoRM
ecosystem to access to the production database.

Initially, the simulation tool, after knowing the available services registered in the ecosystem, queries the Middleware for the availability of production historical data by requesting the execution of the "getData" service
offered by the production database. If a positive response is returned, the
Middleware will trigger the necessary actions for the data retrieval, namely
using a data adapter (in the case where the data source is a legacy). After gathering the requested data, the tool will proceed with the simulation process.

Note also, that this process flow could be made more complex by inserting
multiple queries to different data sources. Additionally, the process could
end by the generation of new data as a result of the simulation process and/
or by triggering further steps; for example, by using scheduling embedded
in the simulation process or scheduling to use the simulation results.

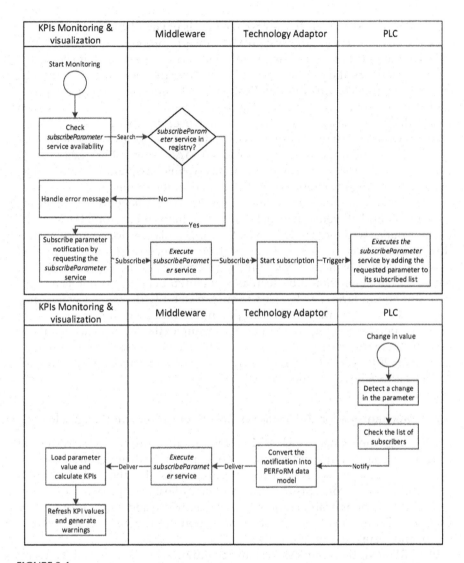

FIGURE 3.4
Exemplary flow diagram for a key performance indicator (KPI) monitoring tool integrated in the Production harmonizEd Reconfiguration of Flexible Robots and Machinery (PERFoRM) ecosystem. (From EU HORIZON 2020 FoF PERFoRM 2016.)

Figure 3.4 depicts a process flow used by the KPI monitoring and visualization tool to get data parameters to support the monitoring of KPIs by using a subscribe/notification procedure. For this purpose, the tool initially subscribes the notification of a parameter change by requesting the *subscribeParameter* service, offered by a PLC located at the shop floor. Since this is a legacy PLC, the request must undertake a data conversion by means of the use of an appropriate technology adapter, converting a non-PERFoRM into a

PERFoRM data structure. In the case the PLC is not available or does not provide the desired service, the system Middleware should reply accordingly.

When the PLC detects a change in the subscribed parameter, it checks the list of subscribers and notifies them of such an occurrence. Naturally, the KPI monitoring and visualization tool can subscribe several parameters, possibly from different sources, according the correlation to calculate the desired KPIs, which are analyzed and displayed in a user dashboard.

As previously referred, these tools use advanced algorithms and technologies, some of them using MAS technology. MAS (Ferber 1999; Leitão 2009) is a suitable approach to provide flexibility, robustness, and responsiveness by decentralizing the control over distributed, autonomous, and cooperative intelligent control nodes. In spite of these important benefits, the real-time constraints and the emergent behavior in industrial environments can be pointed out as weaknesses. However, the MAS tools developed in PERFoRM do not face soft or hard real-time restrictions, since these tools are placed at strategic and/or tactical planning and control levels. Additionally, the emergent behavior should be seen as a potential benefit and not as a problem, since boundaries can be used to ensure stability during the emergency process. In fact, several commercial planning and scheduling solutions are already operating in big companies: for example, the MAS solutions developed by Smart Solutions (2016), as well as the Gartner's Strategic Technology Trends for 2016 report that predicts the use of MAS technology as a base for numerous mobile applications by 2020 (Gartner 2015).

3.2.6 Mechanisms for the Seamless Reconfiguration and Pluggability

All the aforementioned PERFoRM system architecture features would not be fully exploited if the architecture is not enriched with appropriate mechanisms for the seamless system reconfiguration as also the introduction of plug-and-produce concepts for a proper "modularity" approach. The seamless system reconfiguration requires the capability to add, remove, or modify production components on the fly, that is, without the need to stop, reprogram, and restart the components.

In PERFoRM, the seamless system reconfiguration is achieved by using the features commonly used in the development of distributed systems, namely those under the technological umbrella of MAS and SOA, particularly service registry, discovering, and composition, which also enhances the pluggability, and the proper design, development, and deployment of self-mechanisms, especially those targeted for improving the system adaptability and reconfiguration. These technological approaches provide flexibility and robustness associated to their decentralized and distributed nature, in opposite to the traditional centralized control approaches, which are built up upon a central node.

In service-oriented design, entities that want to offer their functionalities, encapsulated as services, should publish these services in a registry

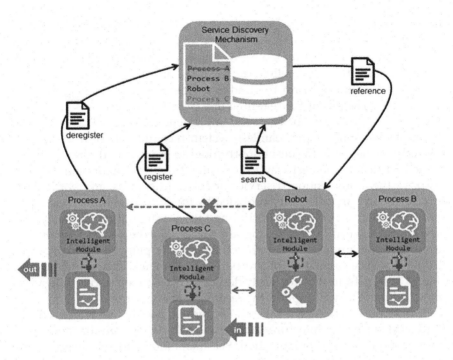

FIGURE 3.5
Service registry and discovering. (From EU HORIZON 2020 FoF PERFoRM 2016.)

repository that acts as a "yellow pages" functionality. In PERFoRM, the hardware devices and software applications need to register themselves into the system and particularly their catalogue of services (i.e., by means of the "yellow pages" registry). The plug-in of new services in the system is easily discovered by the other entities through the use of a service discovering mechanism, potentiating the cooperation/collaboration between different system components leading to the seamless system reconfiguration. For example, as illustrated in Figure 3.5, consider a system comprising the "Process A" and "Process B" that are interacting with an industrial "Robot." In case of system reconfiguration, through replacing the "Process A" by "Process C," the intelligence (e.g., an agent) of the "Process A" should deregister its service from the service registry and the intelligence of the "Process C" should register its service. Automatically, and on the fly, the intelligence of the industrial robot discovers the new service and adapts its internal behavior to start interacting with the new plugged "Process C."

After the discovering process, the services can be requested to be executed in a smoothly manner and following a proper process flow.

It is also worthy to note that in order to have a fully functional system, the service registration must be accompanied with the necessary information regarding the tool that offers them, for example, its location or service quality. Therefore, the service registry and discovering, offered by the Middleware, must follow the best practices of similar MAS and SOA systems already in use. The pluggability of production components, supported by the registry and discovering features, enables the seamless, online, and on-the-fly reconfiguration.

The consideration of intelligence modules attached to the production devices allows to transform the traditional devices into smart production components. The use of the real-time info processing module is one example, but these modules, for example, implemented using ML, can support the implementation of the seamless system reconfiguration. In particular, they can reason about the best opportunities to proceed with the reconfiguration and decide the best way to implement the reconfiguration. For this purpose, the aforementioned service-orientation features, and particularly service registry and discovery provided by the industrial Middleware platform, are used as a mean of performing the reconfiguration on the fly, that is, without the need to stop, reprogram, and restart the system. Usually, the reconfiguration might happen when a resource is plugged-in/out or when the performance parameters of a resource are changed. This reconfiguration "block" differs from the previously described classic tools and is rather a set of procedures that will support the reconfiguration process.

In this perspective, several self-properties can be identified in the PERFoRM system architecture, namely self-adaptation, self-diagnosis, self-optimization, and self-organization. Self-adaptation and self-diagnosis are provided by individual smart production components embedding the real-time info processing modules, and self-optimization is provided by the different advanced tools for planning, scheduling, and simulation running in the PERFoRM cloud environment. The self-organization leading to the system reconfiguration emerges from the interaction of the smart production components that have embedded local and global driving self-organization forces implemented using MAS technology. Note that self-organization plays an important role for the system reconfiguration, namely considering the behavioral and structural perspectives, which provides different scopes and time response to evolution (Barbosa et al. 2015).

Self-properties allow the seamless adaptation and reconfiguration of the system; however, in some circumstances the human role is impacted. In these cases, the operators have to be involved in the reorganization process.

3.3 Positioning the System Architecture within Industry 4.0

The digitalization of the shop floor, as proposed by the PERFoRM project, is aligned with the current state-of-the-art and road-map trends, and particularly follows the major characteristics of the Industry 4.0 platform (Kagermann, Wahlster, and Helbig 2013; ZVEI 2016).

CPSs are at the cornerstone of the Industry 4.0, and PERFoRM addresses this by setting its foundation in the CPS concepts, particularly by promoting the symbiotic use of "digital" and "physical" layers of the manufacturing world and also considering its interconnection and interoperability. Optimized decision-making is also refereed in the report by Kagermann, Wahlster, and Helbig (2013) and PERFoRM is addressing this by promoting a set of different tools that will allow the decision-makers, and particularly each of the use cases, to early detect deviations and performance degradation, allowing to take better, more accurate, and timely decisions. Integration, either vertical and/or horizontal, is also mentioned to be crucial. PERFoRM also addresses this by promoting the use of a common PERFoRM data model, covering data needs from lower levels into higher levels as also from different domains within the same level (e.g., considering different data needs of devices) and by considering a distributed and interoperable Middleware. Finally, and considering only a few key concepts, the Industry 4.0 working group also recommends the development of a reference architecture. This is currently being developed by the Industrie 4.0 platform, named "Reference Architectural Model Industrie 4.0 (RAMI4.0)" (ZVEI 2016), but from the initial developments, PERFoRM is aligned with what is being considered in RAMI4.0 (DIN SPEC 91345:2016-04 2016).

The "connectivity and interoperability" is covered through the development of a distributed and interoperable Middleware alongside with the design of a common and cross-layers data model. The "seamless factory of the future system integration" is accomplished by the connection of several information data sources as also the consideration of the human as a valuable data source itself. The "integration of existent systems" is also managed by the development of hardware and software adapters, adapting the native information language into the PERFoRM ecosystem. Modeling and simulation are also envisioned as crucial building blocks of future systems. Therefore, the PERFoRM architecture considers the use of such tool domains, particularly allowing beforehand to foresee future problems and solutions to these as to allow to optimize production processes. Other key concepts are located around the human operator and in its role in future production systems. PERFoRM considers this by moving the human to the center of the architecture and to consider him as a flexibility driver in future systems. Therefore, a special emphasis is being devoted to the study of its integration and interaction to/from the system.

One of the most critical aspects being pointed out by all the reference documents is the use of standards and the standardization process promotion. PERFoRM also considers this issue as a major road-blocker breaker and enabler for the future industrial adoption of the system architecture, and is promoting the use of standardized approaches and technologies, for example, using OPC-UA or AutomationML data models.

Finally, the smooth, secure, and efficient migration from the traditional centralized structures and legacy systems, currently running in industrial environments, to the emergent distributed, agile, and plug-and-produce systems, requires a special attention (note that newer devices and/or applications will coexist with remaining existing systems). In fact, as also stated in the report by Kagermann, Wahlster, and Helbig (2013), "the journey toward Industrie 4.0 will be an evolutionary process," and also reinforced in the IEC (2015), the migration process from legacy systems is also crucial, particularly in the future adoption of such innovative systems. This issue may be simplified with the definition of migration methodologies and guidelines, and the adoption of industrial Middleware components, standard interfaces, and repository of wrappers, as, for example, described in Chapter 5 of the book (Colombo et al. 2014).

3.4 Conclusion

This chapter describes the design of the modular PERFoRM system architecture, covering all the different layers in the production process identified by ISA-95 automation model, being able to respond in a prompt manner to nowadays requirements as aligned with state-of-the-art visions, such as those advocated by the Industry 4.0 initiative.

The proposed system architecture reused some existing architectural approaches for distributed control systems and complemented with new manufacturing architectural trends. Briefly the main principles of the PERFoRM architecture are as follows:

- The use of service orientation to expose the system functionalities as services.
- The use of a common platform for information exchange, that is, a Middleware.
- The use of a common language for the specification of standard interfaces.
- The compliance with legacy systems by means of technology adapters.
- The use of the human as a flexibility driver.
- The development of advanced planning, simulation, and operational tools.

The designed system architecture was analyzed by its compliance according to its alignment with the Industry 4.0 principles and the major specifications of the RAMI4.0 (IEC 2017).

References

Angione, Giacomo, José Barbosa, Frederik Gosewehr, Paulo Leitão, Daniele Massa, João Matos, et al. 2017. "Integration and deployment of a distributed and pluggable industrial architecture for the PERFoRM project." *Procedia Manufacturing.* 11: 896–904. doi:10.1016/j.promfg.2017.07.193.

Barbosa, J., P. Leitão, E. Adam, and D. Trentesaux. 2015. "Dynamic self-organization in holonic multi-agent manufacturing systems: The ADACOR evolution." *Computers in Industry.* 66: 99–111.

Colombo, A. W., and S. Karnouskos. 2009. "Towards the Factory of the Future: A Service-Oriented Cross-Layer Infrastructure." In *ICT Shaping the World: A Scientific View, European Telecommunications Standards Institute (ETSI),* pp. 65–81. New York: Wiley.

Colombo, Armando W., Ricardo Carelli, and Benjamin Kuchen. 1997. "A temporised Petri net approach for design, modelling and analysis of flexible production systems." *Int. J. Adv. Manuf. Technol.* 13 (3): 214–226. doi:10.1007/BF01305873.

Colombo, Armando W., Thomas Bangemann, Stamatis Karnouskos, Jerker Delsing, Petr Stluka, Robert Harrison, et al. 2014. *Industrial Cloud-Based Cyber-Physical Systems.* Cham: Springer International Publishing.

Colombo, Armando W., Ronald Schoop, Ralf Neubert. 2004. "Collaborative (agent-based) factory automation". The Industrial Information Technology Handbook. CRC Press, pp. 1759–1779.

DIN SPEC 91345:2016-04. 2016. "Reference Architecture Model Industrie 4.0 (RAMI4.0): Beuth Verlag GmbH." https://www.beuth.de/en/technical-rule/din-spec-91345-en/250940128.

EU HORIZON 2020 FoF PERFoRM. 2015. "Deliverable D2.1: Guidelines for Seamless Integration of Humans as Flexibility Driver in Flexible Production Systems." Unpublished manuscript, last modified 10, 2018. http://www.horizon2020-perform.eu/index.php?action=documents.

EU HORIZON 2020 FoF PERFoRM. 2016. "Deliverable D2.2: Definition of the System Architecture." Unpublished manuscript, last modified 10, 2018. http://www.horizon2020-perform.eu/index.php?action=documents.

EU HORIZON 2020 FoF PERFoRM. 2017. "Deliverable D2.5: Guidelines for seamless integration of Humans as flexibility driver in flexible production systems." Unpublished manuscript, last modified 10, 2018. http://www.horizon2020-perform.eu/index.php?action=documents.

Feldmann, K., and A. W. Colombo. 1998. "Material flow and control sequence specification of flexible production systems using coloured Petri nets." *Int. J. Adv. Manuf. Technol.* 14 (10): 760–774. doi:10.1007/BF01438228.

Ferber, J. 1999. *Multi-Agent Systems: An Introduction to Distributed Artificial Intelligence.* Addison-Wesley Longman. Harlow, UK.

Gartner. 2015. "Gartner Identifies the Top 10 Strategic Technology Trends for 2016." Accessed June 30, 2016. http://www.gartner.com/newsroom/id/3143521/.

Gosewehr, Frederik, Jeffrey Wermann, Waldemar Borsych, and Armando W. Colombo. 2017. "Specification and design of an industrial manufacturing middleware." Proc. of the 15th IEEE Int. Conf. on Industrial Informatics INDIN2017, pp. 1160–1166.

Haskamp, Hermann, Michael Meyer, Romina Mollmann, Florian Orth, and Armando W. Colombo. 2017. "Benchmarking of existing OPC UA implementations for industrie 4.0-compliant digitalization solutions." Proc. of the 15th IEEE Int. Conf. on Industrial Informatics INDIN2017, pp. 589–594.

IEC. 2015. "White paper: Factory of the future." *International Electrotechnical Commission*. ISBN 978-2-8322-2811-1.

IEC. 2017. "Smart manufacturing - Reference architecture model industry 4.0 (RAMI4.0): 1.0th ed. 25.040.01 - Industrial automation systems in general 35.080 - Software 35.240.50 - IT applications in industry, no. 63088:2017: IEC." https://webstore.iec.ch/publication/30082.

ISO/IEC/IEEE 42010. 2011. "Systems and software engineering - Architecture description." (n.d.). Accessed August 28, 2018. https://www.iso.org/standard/50508.html.

Jammes, F., H. Smit, J. L. Lastra, and I. Delamer. 2005. "Orchestration of Service-oriented Manufacturing Processes." *Proceedings of the 10th IEEE International Conference (ETFA'05)*, 617–624.

Kagermann, H., W. Wahlster, and J. Helbig. 2013. "Securing the Future of German Manufacturing Industry: Recommendations for Implementing the Strategic Initiative INDUSTRIE 4.0." Technical report, ACATECH – German National Academy of Science and Engineering.

Leitão, P. 2009. "Agent-based distributed manufacturing control: A state-of-the-art survey." *Engineering Applications of Artificial Intelligence*. 22: 979–991.

Leitão, P., J. Barbosa, A. Pereira, J. Barata, and A. W. Colombo. 2016. "Specification of the PERFoRM Architecture for Seamless Production System Reconfiguration." *Proceedings of the 42nd Annual Conference of IEEE Industrial Electronics Society (IECON'16)*, October 24–27, Firenze, Italy, 5729–5734. doi: 10.1109/IECON.2016.7793007.

Mendes, J. M., A. Bepperling, J. Pinto, P. Leitão, F. Restivo, and A. W. Colombo. 2009. "Software Methodologies for the Engineering of Service-oriented Industrial Automation: The Continuum Project." *Proceedings of the 33rd Computer Software and Applications Conference (COMPSAC'09)*, 452–459.

Murata, T. 1989. "Petri nets: Properties, analysis and applications." *IEEE*. 77 (4): 541–580.

Peltz, C. 2003. *Web Services Orchestration and Choreography*. IEEE Computer, Vol. 36, Issue 10.

Smart Solutions. Accessed June 30, 2016. http://smartsolutions-123.ru/en/.

"SOCRADES." http://www.socrades.eu/.

The European Factories of the Future Research Association. 2013. "Factories of the Future Roadmap." Accessed October 10, 2018. https://www.effra.eu/factories-future-roadmap.

ZVEI. 2016. "Industrie 4.0: The Reference Architectural Model Industrie 4.0 (RAMI 4.0)." Accessed June 30, 2016. http://www.zvei.org/Downloads/Automation/ZVEI-Industrie-40-RAMI-40-English.pdf.

Wooldridge, M. 2002. "An Introduction to MultiAgent Systems", 2nd Edition. John Wiley and Sons, Chichester, UK.

4

Architectural Elements: PERFoRM Data Model

Ricardo Silva Peres, Andre Dionisio Rocha, José Barata
(UNINOVA - Instituto de Desenvolvimento de Novas Tecnologias)

Armando W. Colombo
(Institute for Informatics, Automation and Robotics (I²AR))

CONTENTS

4.1 Introduction

With the emergence of the Industry 4.0 paradigm, the industry is bearing witness to the appearance of more and more complex systems, often requiring the integration, both connectivity and interoperability, of various new heterogeneous, modular, and intelligent elements (digitalized assets, things) with preexisting legacy devices and systems.

This challenge of interoperability is one of the main concerns taken into account when designing such industrial systems-of-cyber-physical systems, commonly requiring the use of standard interfaces to ensure this seamless integration (Colombo et al. 2017). To aid in tackling this challenge, a common format for data representation, as well as standard interfaces for data exchange, interoperability, and service exposure should be adopted.

Aligned with PERFoRM's Industry 4.0 vision, this chapter details the design of a common data model and two standard interfaces, which can be used as some of the core elements to enable the seamless integration and interconnectivity between both new digitalized and harmonized with legacy systems, taking into account the specific needs of four different use cases representing varied European industry sectors.

4.2 Definition

In the context of the PERFoRM approach, the data model and standard interface are a set of definitions that assist in the connection of its varied heterogeneous elements as well as in the representation of legacy production systems to be abstracted.

The main goal for the data model is to provide a common way through which data and assets can be represented and exchanged between the different components comprised in the PERFoRM solution. In line with this, the standard interface provides a way for PERFoRM elements to interact with the Middleware in a transparent manner through a set of standard calls.

4.3 Role within PERFoRM

An important key issue to ensure the interoperability in real industrial environments, interconnecting heterogeneous legacy hardware devices (e.g., robots and Programmable Logic Controllers [PLCs]) and software applications (e.g., Supervisory Control and Data Acquisition [SCADA], the Manufacturing Execution System [MES], and databases) is the adoption of standard interfaces.

These aim to define the bridge between devices and applications in a unique, standard, and transparent manner, ensuring the transparent pluggability of these heterogeneous devices. For this purpose, a standard data representation should be adopted by the interface that should also define the list of services provided by it, and the semantics data model handled by each service.

In this definition, and particularly for industrial automation, several ISA-95 layers addressing different data scope and requirements should be considered, namely the machinery level covering mainly L1 (automation control) and L2 (supervisory control) layers, and the backbone level covering the L3 (manufacturing operations management) and L4 (business planning and logistics) layers.

Additionally, the achievement of complete interoperability and pluggability requires to complement the use of a standard interface with adapters to transform the legacy data representation into the native standard interface data format.

The validation of the PERFoRM system has been accomplished in four use cases, covering a wide spectrum of the European industrial force, ranging from home appliances to aerospace and from micro electrical vehicles to large compressor production (see Chapters 10–13). Several requirements were collected from the use cases, which were then clustered to be considered in the specification of the system architecture, namely focusing the flexibility, reconfigurability, and general issues (see Table 4.1).

The flexibility cluster identifies a set of requirements related to the ability to change production processes in an agile manner, and to adapt the cycle times and their associated costs, namely the ability to change raw material, ability to change processes, ability to obtain process interactions, agility of production, easy mobility, reduction of the cycle time, and reduction of the cycle cost.

The second cluster considers requirements related to the reconfigurability aspect, namely the need to have several feedback loops between the different phases of the production process and a decrease of setup times due to the system reconfiguration. Additionally, there is the need to have the possibility to integrate the robot programming with the technical Computer-Aided

TABLE 4.1

General Requirements: Flexibility and Reconfigurability Overview

General Requirements		Other Requirements
Flexibility	**Reconfigurability**	**Necessary to Flexibility and Reconfigurability**
• Ability to change raw material • Ability to change processes • Ability to obtain process interactions • Agility production • To facilitate mobility, including comparison among different units, e.g., Overall Equipment Effectiveness (OEE), micro-flow cells • Cycle time reduction • Cycle, cost reduction	• To obtain feedback from production to design • To obtain final test feedback to Robot system configuration • To obtain feedback to the process, based on failure control • Cost saving in reconfiguration To obtain new part reprogramming/set up through Computer-Aided Design (CAD) critical paths: (i) Self-configuring system, which can define the root cause based on pattern recognition; (ii) Setup time reduction	• 100% traceability and identification of single products up to the supply chain • Ability to enable simulation, model and prototype in the Cyber-Physical Systems (CPSs) environment (i.e., process parameters interaction, global factory behavior, predictive failure) • An increase in the amount of data collected and data availability • Automatic (semiautomatic) data gathering of machine condition • Full integration and quick communication among different departments and functions (i.e., scheduling system and maintenance system integration, machine condition and maintenance tasks, production and process planning, etc.)

Design (CAD) drawings, obtain feedback from production to design, obtain final test feedback to the robot system configuration, obtain feedback to the process based on failure control, cost saving in reconfiguration, obtain new part reprogramming/set up through CAD critical paths, self-configuring system, and reduction of the setup time.

The *General Requirements* include only the requirements regarding flexibility and reconfigurability that are required across all four use cases, while the *Other Requirements* column lists requirements that are specific to certain use cases and that lead to obtain those aspects of flexibility and reconfigurability grouped in the *General Requirements*.

It is observable that product and production traceability is mandatory, as well as the automatic gathering of data and the use of simulation tools in the process chain (DIN/VDE 2017). Furthermore, the integration of systems from different company's departments is of major concern. To accomplish the requirements, the architecture proposed in Chapter 3 identifies the need of developing a common data model and generic standard interfaces. In fact, and in line with the general vision for the Industry 4.0 platform (Platform i40 2018), one of the key challenges to tackle the requirements addressing flexibility and reconfigurability is the aspect of interoperability in real industrial environments, dealing with the representation and seamless exchange of data originating from a wide array of entities, often from very different, albeit related, actions levels. The interconnection of heterogeneous legacy HW devices (e.g., robots and the respective controllers) and SW applications (e.g., databases, SCADA applications, and other management, analytics, and logistics tools) is one of the main goals currently being pursued in this vision.

To this effect, the PERFoRM architecture employs the adoption of a standard interface as the main driver for pluggability and interoperability, aiming at enabling the connection between such devices and applications in a seamless and transparent manner. This interface should provide the devices, tools, and applications with the means to fully expose and describe their services in a unique, standardized, and transparent way to enhance the seamless interoperability and pluggability, fully specifying the semantics and data flow involved in terms of inputs and outputs required to interact with these elements.

Therefore, it should provide a set of functionalities related to a standardized service invocation, that is, the definition of the list of services to be implemented by the interface, the contract implementation of each service (i.e., the name, input parameters, and output parameters), and the definition of the data model handled by the services. Note that an important requirement for the design of the standard interfaces is the usage of service orientation to expose the device/system functionalities as services.

For this purpose, a common data model is also adopted, serving as the data exchange format shared between the PERFoRM-compliant architectural elements. This data model needs to cover the semantic needs associated with each entity, which, in the particular case of industrial automation, means that the requirements related to each of the ISA-95 layers and their respective needs are considered.

The specification and implementation details of the data model and standard interface will be described in Section 4.

4.4 Implementation Details

This section entails the description of PERFoRMML, which is PERFoRM's common data model and its standard interface. The implementation of PERFoRMML was conducted in AutomationML taking into consideration all the requirements from the different use cases and tool developers.

4.4.1 PERFoRM Data Model

The overall view of the class diagram can be seen in Figure 4.1.

One of the most important facets is that of machinery and control systems, which encompasses all the elements necessary to model the system's topology, data types, and interaction at physical machinery level. A few examples are provided below.

The *PMLParameter* and *PMLValue* elements enable the basic representation of information pertaining to data at the machinery level, namely in terms of parameters for configurations and skills (abilities, functions, or tasks performed by shop floor elements), and shop floor data to be extracted from various sources such as PLCs and databases, respectively.

The *PMLEntity* class is the generic representation of shop floor entities, encapsulating all the necessary information that is associated simultaneously to both components and subsystems (set of components and possibly other subsystems working toward a common goal).

Additionally, despite not being directly used, the *PMLEntity* class enables the existence of generic collections of elements that can be either components or subsystems, without *a priori* knowledge of what the composition of such collections will be.

In order to abstract components and subsystems in the shop floor, both the *PMLComponent* and *PMLSubsystem* classes extend the characteristics of a *PMLEntity*, as it can be observed in Figure 4.1.

The former indicates the finest level of granularity, therefore representing a single component in the shop floor which may offer a given number of skills (e.g., pick, place, move, weld) and may possess certain values that are relevant to be extracted (e.g., cycle time, energy consumption, sensor data). This representation is generic enough in order for components to be able to refer not only to physical machine resources, but also to virtual ones

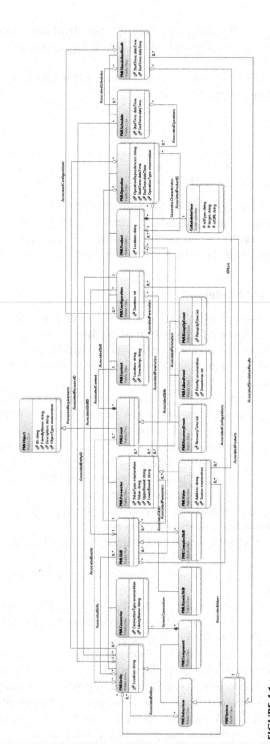

FIGURE 4.1
PERFoRMML class diagram. (From EU HORIZON 2020 FoF PERFoRM 2016.)

or even human operators. As such, a *PMLComponent* is essentially a single entity that can perform skills and present relevant data regarding its state and operation.

In turn, the *PMLSubsystem* provides similar functionality, albeit regarding subsystems instead, thus referring to a group of components and possibly other subsystems. It is a recursive element in the sense that it extends the *PMLEntity* class, while also being able to contain other *PMLEntities* within.

Furthermore, the generic design allows different levels of granularity to be targeted using the same data model. A system can be modeled in such a way that a robot is the lowest entity in terms of the abstraction level, being regarded as a simple component, while that same robot can be seen as a subsystem within the system itself, encompassing several sensors as its components, depending on the desired level of granularity.

These classes were implemented using AutomationML (AutomationML-Organization 2018). Based on a study assessing several data representation and exchange formats (Peres et al. 2016), it was chosen as the most suitable to meet PERFoRM's requirements. It consists of a neutral data format based on XML for the storage and exchange of plant engineering information using a combination of different standards. It can thus be used to connect the heterogeneous tools that make up the PERFoRM solution, across their wide array of disciplines.

In Figure 4.2, an excerpt of the resulting System Unit Class library can be observed.

These constitute the group of reusable classes that can be used to create the instance model of a specific use case. These classes are the core of PERFoRM's data model implementation, being the direct translation of the model represented in Figure 4.1.

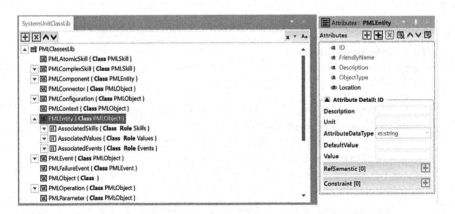

FIGURE 4.2
AutomationML SystemUnitClassLib.

4.4.2 Standard Interfaces

As previously mentioned, one of the key aspects for tackling PERFoRM's interoperability challenges, mainly in regard to the seamless exchange of data between heterogeneous entities, is the adoption of a standard interface. This acts as the main driver for pluggability and interoperability, enabling the interconnection of both software and hardware entities across the different manufacturing layers in a seamless and transparent fashion.

As such, full interoperability and harmonization of data at a system-of-systems level are achieved by coupling the standard interface with the data model for a common representation of data and system semantics. However, taking into account the integration of legacy devices and their own individual data models and semantic requirements, the addition of technology adapters is also required in order to enable the translation and mapping of legacy data into the common PERFoRM representation, allowing for these devices to be conferred additional intelligence and integrated into the cyber-physical paradigm.

The backbone interface is responsible for exposing the functionalities of the tools connected to the Middleware, as well as allowing these tools to communicate with the lower levels in order to acquire any information required to execute their respective tasks. An overview of this interface can be found in Figure 4.3.

This interface enables PERFoRM tools to interact in a generic way using two commonly used approaches. The first two methods, *GetPMLObjects* and *SetPMLObjects* allow a tool to either request or send a particular *PMLObject*, akin to for instance a typical REST API, simply by indicating the type of object being exchanged and the respective endpoint. The remaining two enable tools to use a publish and subscribe approach, where tools can for instance subscribe to a particular topic (e.g., simulation results) and receive new data as it becomes available from a different tool.

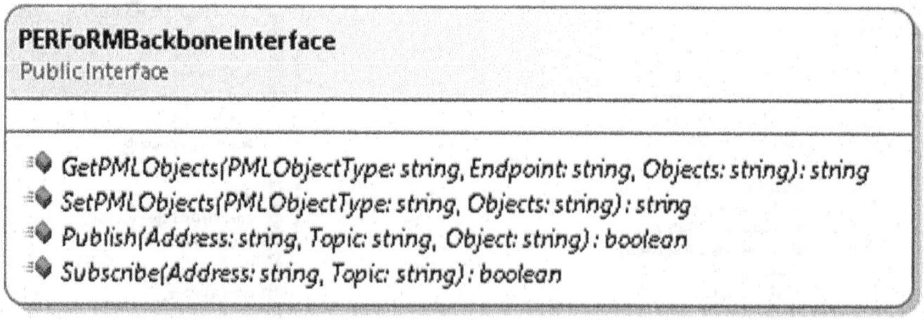

```
PERFoRMBackboneInterface
Public Interface

◆ GetPMLObjects(PMLObjectType: string, Endpoint: string, Objects: string): string
◆ SetPMLObjects(PMLObjectType: string, Objects: string): string
◆ Publish(Address: string, Topic: string, Object: string): boolean
◆ Subscribe(Address: string, Topic: string): boolean
```

FIGURE 4.3
PERFoRM Backbone Interface. (From EU HORIZON 2020 FoF PERFoRM 2016.)

4.5 Example

A major aspect is the harmonization of the efforts between the developments of the architecture, the data model and interfaces, the Middleware and the adapters (Angione et al. 2017). In order to optimally achieve that harmonization, a test case focused on a punching cell was selected and implemented. It is mainly consisting in a conveyor belt and its DC motor (*M*1), a punching cylinder and respective DC motor (*M*2), two-part presence sensors (*S*1 and *S*2, phototransistors), and two switches (*A*1 and *A*2), as illustrated in Figure 4.4.

The relevant data available to be extracted from the cell is described in Table 4.2.

The punching process is relatively simple, consisting essentially in a part entering the cell at *S*1, being then moved to *S*2 in order for the motor *M*2 to

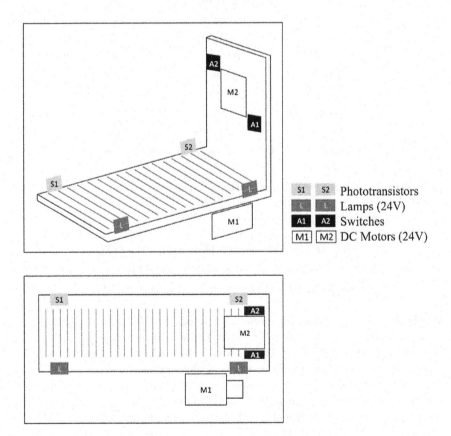

FIGURE 4.4
Punching cell overview. (From EU HORIZON 2020 FoF PERFoRM 2016.)

TABLE 4.2

Punching Cell Values

Entity	Value	Source
M1	M1F	PLC
M1	M1T	PLC
M2	M2D	PLC
M2	M2U	PLC
M1	S1	PLC
M1	S2	PLC
M2	A1	PLC
M2	A2	PLC
M1	Humidity	DB
M1	Temperature	DB
M1	HeatIndex	DB
M1	BatteryVoltage	DB
M1	Humidity	DB
M2	Temperature	DB
M2	Pressure	DB
Cell	InExecution	PLC
Cell	ProcessingTime	PLC

Source: EU HORIZON2020 FoF PERFoRM 2015-2018.

initiate the punching of said part. Afterwards, the part is simply moved back along the conveyor (controlled by $M1$) to $S1$.

The process flow works as follows:

$$S1\,(1) > M1F\,(1) > S1\,(0) > S2\,(1) > M1F\,(0) > M2D\,(1) >$$
$$A2\,(0) > A1\,(1) > M2D\,(0) > M2U\,(1) > A1\,(0) > \qquad (4.1)$$
$$A2\,(1) > M2U\,(0 > M1T\,(1) > S2\,(0) > S1\,(1) > M1T\,(0)$$

While the PLC provides the low-level sensor and I/O data regarding each of the components, as shown in Table 4.2, a database also provides some additional information for both motors, mainly regarding readings such as temperature, humidity, voltages, and hydraulic pressure.

For the harmonization's purposes, the granularity target was defined at the motor level, being clear that these two entities, $M1$ and $M2$, provide the four main driving skills for the cell, namely the conveyor control, $M1F$ and $M1T$, and the punching control, $M2U$ and $M2D$, respectively.

As such, taking into account the relevant values extracted from Table 4.2, as well as the overall topological organization of the cell, the following PERFoRMML model was obtained, as shown in Figure 4.5.

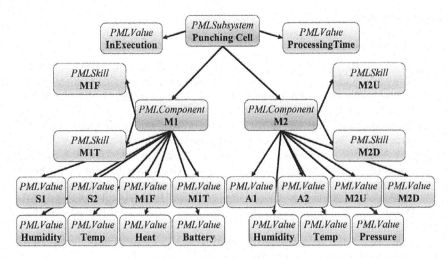

FIGURE 4.5
Punching cell modeling. (From EU HORIZON2020 FoF PERFoRM 2015-2018)

4.6 Conclusion

In summary, the present chapter encompassed the description of both PERFoRM's common data model and standard interface. These elements serve as one of the main drivers for seamless interconnectivity and data exchange between the heterogeneous elements that make up the PERFoRM ecosystem.

For this purpose a brief discussion about the role of these architectural elements within the PERFoRM approach was presented along with an overview of the general and specific requirements that guided their development. This process ensures that not only the expectations of the ecosystem's tool developers are met, but also those of the industrial partners, thus promoting a well-established alignment with project's end users. Furthermore, the details of the data model implementation were provided, starting from the inception of its design all the way up to its realization using AutomationML (AutomationML-Organization 2018), a data exchange format aligned with the Reference Architectural Model for Industry 4.0 (RAMI 4.0) (DIN 91345:2016-04 2016).

Finally, the usage of the developed elements and their integration with the remaining constituents of the PERFoRM ecosystem was illustrated using the example from an integration use case. In this example, a punching cell was modeled consisting in a conveyor belt and its DC motor, a punching cylinder and respective DC motor, two-part presence sensors and two switches. This effectively showcased the generic applicability of the PERFoRM data model and its role in the harmonization of the varied elements involved in the solution.

References

Angione, G., J. Barbosa, F. Gosewehr, P. Leitão, D. Massa, J. Matos, et al. 2017. "Integration and deployment of a distributed and pluggable industrial architecture for the PERFoRM project." *Procedia Manufacturing.* 11: 896–904. https://doi.org/10.1016/j.promfg.2017.07.193.

AutomationML-Organization. 2018. "AutomationML." Accessed October 12, 2018. https://www.automationml.org/o.red.c/home.html.

Colombo, Armando W., Stamatis Karnouskos, Okyay Kaynak, Yang Shi, and Shen Yin. 2017. "Industrial cyberphysical systems: A backbone of the fourth industrial revolution." *IEEE Ind. Electron. Mag.* 11 (1): 6–16. doi:10.1109/MIE.2017.2648857.

DIN 91345:2016-04. 2016. *Reference Architecture Model Industrie 4.0 (RAMI4.0): Beuth Verlag GmbH.* https://www.beuth.de/en/technical-rule/din-spec-91345-en/250940128.

DIN/VDE. 2017. "DIN EN 62890:2017-04." Accessed October 10, 2018. https://www.beuth.de/de/norm-entwurf/din-en-62890/269121145.

EU HORIZON 2020 FoF PERFoRM. 2015-2018. *"Deliverable D2.3: Specification of the Generic Interfaces for Machinery, Control Systems and Data Backbone."* Unpublished manuscript, last modified October 10, 2018. http://www.horizon2020-perform.eu/index.php?action=documents.

Peres, R. S., M. Parreira-Rocha, A. D. Rocha, J. Barbosa, P. Leitão, and J. Barata. 2016. "Selection of a Data Exchange Format for Industry 4.0 Manufacturing Systems." In *IECON 2016—42nd Annual Conference of the IEEE Industrial Electronics Society,* pp. 5723–5728. IEEE. https://doi.org/10.1109/IECON.2016.7793750.

Platform i40. 2018. *"Industrie4.0."* Accessed October 12, 2018. https://www.plattform-i40.de/I40/Navigation/DE/Home/home.html.

5

Architectural Elements:
Technology Adapters

Cristina Cristalli, Giacomo Angione, Giulia Lo Duca

(Loccioni)

Martin Frauenfelder, Simon Eggimann

(Paro AG)

CONTENTS

5.1 Introduction

With the arrival of fourth Industrial Revolution and the propagation of Industry 4.0 concepts, the complexity of industrial systems is rapidly increasing, often requiring the integration of various new heterogeneous, modular, and intelligent elements with preexisting legacy devices.

Interoperability is one of the key challenges when designing such systems-of-systems, coping with the representation, and seamless exchange of data originating from a wide array of entities, often from different, albeit related, actions levels. Therefore, the interconnection of heterogeneous legacy hardware devices (e.g., robots and the respective controllers) and software applications (e.g., databases, Supervisory Control and Data Acquisition [SCADA] applications, and other management, analytics, and logistics tools) has been one of the goals pursued in the PERFoRM project (EU HORIZON2020 FoF PERFoRM 2015-2018).

As described in this chapter, in order to facilitate the integration of legacy hardware and software production systems, a series of customized Adapters have been developed. The main role of these Adapters is to translate the device/module internal data model into the one defined in the PERFoRM approach.

5.2 Definition

In the PERFoRM approach, the Technology Adapters are hardware and software components that facilitate the connection of production legacy systems to the PERFoRM Middleware component.

The main objective of the Adapters is to transform the legacy data model into the standard interface data model defined in the project (i.e., masking the legacy systems' data/functionalities according to the PERFoRM standard interfaces).

5.3 Role within PERFoRM

The innovative architecture proposed in the PERFoRM project and described in Chapter 3 can be industrially accepted and really adopted only if it is possible to integrate legacy systems. For this reason, technology Adapters are key elements to connect legacy systems (heterogeneous hardware devices and software applications) to the PERFoRM Middleware component and to transform the legacy data model into the standard interface data model.

The PERFoRM architecture (Figure 5.1) exploits the usage of standard interfaces throughout the whole system as the main drivers for pluggability and interoperability, aiming at enabling the connection between such devices and applications in a seamless and transparent manner. These interfaces support devices, tools, and applications with the means to fully expose and describe their services in a unique, standardized, and transparent way to enhance the seamless interoperability and pluggability. Additionally, and as a support to the standard interfaces, a common data model is adopted, serving as the data exchange format shared between the PERFoRM-compliant architectural elements. This data model covers the semantic needs associated to each entity, namely the requirements related to each ISA-95 layer (ISA Organization 2018).

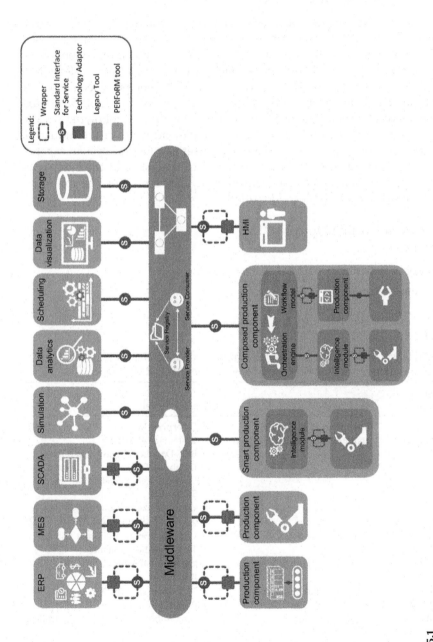

FIGURE 5.1

Adapters in the Production harmonizEd Reconfiguration of Flexible Robots and Machinery (PERFoRM) system architecture. (From EU HORIZON 2020 FoF PERFoRM 2016.)

5.4 Overview about the Implemented Adapters

The application of the PERFoRM approach has been addressing four industrial scenarios with different reconfiguration objectives: predictive maintenance, production automation, Key Performance Indicator (KPI) real-time monitoring, and robot programming. For each scenario, a list of the legacy systems that need to be integrated in the PERFoRM ecosystem has been provided (Table 5.1).

In order to decide which of the legacy systems reported in Table 5.1 should be considered for the implementation of the Adapters, the following parameters have been taken into consideration by the end users of the industrial scenarios: cost effectiveness, obsolesce risk, user satisfaction, system capability, and system integration. According to this analysis and the typologies of the legacy systems considered in the industrial scenarios, the following classes of Adapters have been identified:

- Database (DB) Adapter
- PLC Adapter
- Robot Adapter
- Sensor Adapter

TABLE 5.1

Industrial Scenarios and Legacy Systems in the Production harmonizEd Reconfiguration of Flexible Robots and Machinery (PERFoRM) Project

Industrial Scenario	Objective	Legacy Systems
Siemens Compressors	Integration of a predictive maintenance system	• Enterprise Ressource Planning (ERP) System (SAP APO) • BDE Data Logging System (Oracle Database [DB]) • LHnet Ticketing System (MS SQL Server DB) • Computerized Numerical Control (CNC) Machines (SINUMERIK 840D)
IFEVs Micro-Electrical Vehicles	Automation of the production line	• Welding Robotic Cells and Powertrain Testing Stations (Siemens programmable logic controller (PLC) IM-151)
Whirlpool Microwave Ovens	• Implementation of a key performance indicator (KPI) real-time monitoring system • Reconfiguration of the path of the robot for the leak test	• PERFoRM DB (MS SQL Server DB) • PLM Repository (txt file) • Leak Robot Station (UR10 Controller)
GKN Turbine Vanes	Construction of a reconfigurable robotic cell	• Robotic Cell PLC • Roughness Process (Mitutoyo SJ-210)

TABLE 5.2

Legacy Systems and Technology Adapters Classes

Industrial Scenario	Legacy System	Technology Adapters			
		DB Adapter	PLC Adapter	Robot Adapter	Sensor Adapter
Siemens Compressors	EPR System (SAP APO)				
	Betriebsdatenerfassung (BDE) (Eng. operational data collection) Data Logging System (Oracle DB)	X			
	LHnet Ticketing System (MS SQL Server DB)	X			
	CNC Machines (SINUMERIK 840D)				
IFEVs Micro-Electrical Vehicles	Welding Robotic Cells and Powertrain Testing Stations (Siemens PLC IM-151)		X		
Whirlpool Microwave Ovens	PERFoRM DB (MS SQL Server DB)	X			
	Product Lifecycle Management (PLM) Repository (txt file)				
	Leak Robot Station (UR10 Controller)			X	
GKN Turbine Vanes	Robotic Cell PLC		X		
	Roughness Process (Mitutoyo SJ-210)				X

As shown in Table 5.2, these classes cover 70% of the legacy systems considered for the Adapters implementation and fulfill the objective of the project to implement at least three different production resources Adapters.

5.5 Examples

The following sections describe the results of the implementation of the technology Adapters for each class.

5.5.1 DB Adapter

The DB Adapter permits to connect a legacy DB to the PERFoRM Middleware component. In the PERFoRM approach two types of databases are used in the scenarios of the end users: Oracle for the Siemens use case and MS SQL Server for both use cases, that is, Siemens (see Chapter 10) and Whirlpool (see Chapter 11). The main difference between the two Relational Database Management System (RDBMS) is the language they use. Although both systems use a version of SQL, MS SQL Server uses Transact-SQL (T-SQL), which is an extension of SQL used by Microsoft; Oracle, meanwhile, uses Procedural Language/SQL (PL/SQL).

Typically, software applications are written in a specific programming language (such as Java, C#, etc.), while databases accept queries in some other DB specific language (such as SQL). Therefore, when a software application needs to access data in a DB, an interface that can translate languages to each other (application and DB) is required. Otherwise, application programmers need to learn and incorporate DB specific languages within their applications. Open Database Connectivity (ODBC) and Java Database Connectivity (JDBC) are two interfaces that solve this specific problem. ODBC is a platform, language, and operating system independent interface that can be used for this purpose. Similarly, JDBC is a data Application Programming Interface (API) for the Java programming language. Java programmers can use JDBC-to-ODBC bridge to talk to any ODBC-compliant DB.

Since the PERFoRM Middleware component (Chapter 6) uses Apache ServiceMix (http://servicemix.apache.org/), which is Java-based, the JDBC interface is used for the implementation of the DB Adapter. JDBC Architecture consists of two layers:

- JDBC API: This provides the application-to-JDBC Manager Connection.
- JDBC Driver API: This supports the JDBC Manager-to-Driver Connection.

The JDBC API uses a driver manager and DB-specific drivers to provide transparent connectivity to heterogeneous databases. The JDBC driver manager ensures that the correct driver is used to access each data source. The driver manager is capable of supporting multiple concurrent drivers connected to multiple heterogeneous databases.

In particular, the DB Adapter is able to perform the following steps (Angione et al. 2017):

1. Implement the PERFoRM's standard interface and methods.
2. Connect to the DB (instantiating a DriverManager object).
3. Send queries and update statements to the DB (instantiating statement objects).
4. Retrieve the results from the DB (instantiating ResultSet objects).
5. Present the results according to the PERFoRMML data model (Chapter 4).

According to the PERFoRM architecture (Chapter 3), DB Adapter is implemented as a series of RESTful web APIs (one for each query) which, in the specific industrial scenario, are capable of retrieving data about, for example, production alarms, tasks, orders, etc. Following, some examples of URLs created:

- http://host:port/services/data/production/alarms/{machine}/ {start}/{end}
- http://host:port/services/data/production/tasks/{machine}/{start}/ {end}
- http://host:port/services/data/production/orders/{start}/{end}

When a GET method is received, the Adapter serializes the results of the query in a JavaScript Object Notation (JSON) string according to the PERFoRMML data model (Chapter 4).

5.5.2 Robot Adapter

The Robot Adapter permits to connect a legacy robot to the PERFoRM Middleware (Chapter 6). In particular, the implementation considers the requirements coming from the Leak Robot Station in the Whirlpool use case. In this scenario, a robot (UR10) is equipped with a probe able to detect microwaves leakages. The robot moves the probe following a predefined path around the microwave oven. If leakages are detected the oven is send for reparation.

In this use case, the role of the Adapter is to automatically generate the robot commands starting from a path drawn with a computer-aided design (CAD) tool. This permits to Whirlpool product designers to directly draw the path which is more appropriate for detecting possible microwave leakages outside the oven. Moreover, it allows not to stop the production line when a new model of oven is being produced for teaching the new path to the robot.

After the design of the robot path, the CAD tool exports the three-dimensional (3-D) model into the COLLAborative Design Activity (COLLADA) file format which is part of AutomationML and compliant with PERFoRMML data model (Chapter 4).

Figure 5.2 shows the Graphical User Interface (GUI) of the Robot Adapter developed using LabVIEW programming environment. The 3-D picture displays the 3-D model contained in the COLLADA file, while the controls permit to set the Tool Center Point (TCP) and the acceleration and the speed of the tool. The TCP represents the position of the probe respect to the manipulator turning disk.

The Robot Adapter has been deeply tested at Manufacturing Technology Centre (MTC) laboratory (see MTC 2018) for validating its functionalities before the deployment at the end-user's site (Whirlpool factory in Cassinetta, Italy) (Chakravorti et al. 2017). In particular, in order to replicate the decoupling among the individual systems in the demonstration scenario, the software architecture depicted in Figure 5.3 was set up.

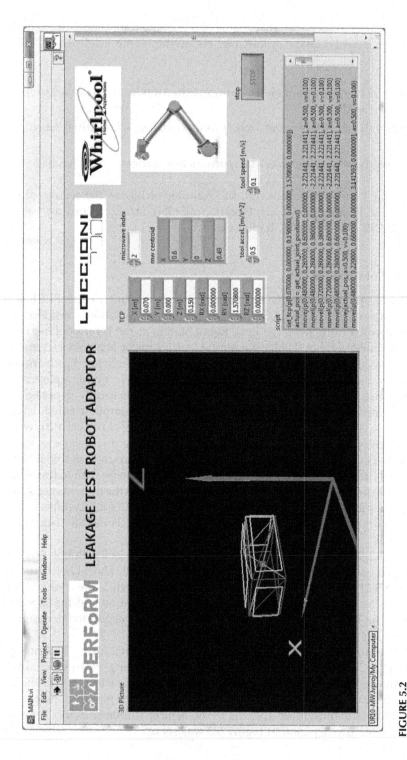

FIGURE 5.2
Robot Adapter Graphical User Interface (GUI). (From EU HORIZON 2020 FoF PERFoRM 2016.)

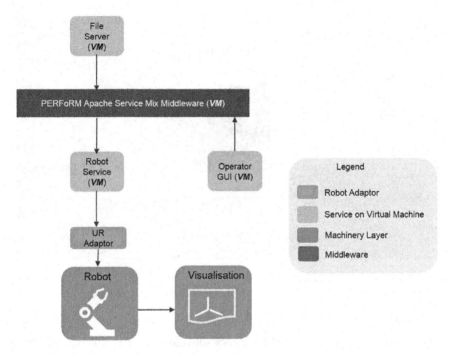

FIGURE 5.3
Software architecture of the Robot Adapter demonstration scenario. (From Chakravorti et al. 2017.)

The demonstration scenario consists of a UR10 robot, the Robot Adapter and four Virtual Machines (VM) with the following functions:

- File Server VM: Hosts the CAD files for performing the robot leak test.
- Robot Service VM: Is able to fetch a particular CAD file from the File Server and push it to a folder that the UR Adapter can access.
- Operator GUI VM: Reads the list of CAD files from the file storage location, displays them, and gives an operator the option of running a particular file.
- Middleware VM: Hosts PERFoRM's Middleware architecture that comprises of an Apache ServiceMix.

In the demonstration scenario, the CAD file is stored on the File Server, the Robot service fetches the CAD file and places it in a specific folder through the PERFoRM Apache ServiceMix Middleware. The Robot Adapter is then able to read the file folder, fetch the file, and translate it to URscript. Finally, the URscript is sent to the robot, which executes the path contained in the CAD file.

FIGURE 5.4
Robot Adapter demonstration scenario in the Manufacturing Technology Centre (MTC) laboratory.

For testing purposes, a software tool (developed by the MTC) has been used to record the linear positions and rotation coordinates of the robot. The route that the robot has followed is compared to the route that it was supposed to follow and the results are visualized by a Matlab script.

Figure 5.4 shows a photo of the demonstration scenario in the MTC laboratory.

5.5.3 Sensor Adapter

The Sensor Adapter permits to connect a legacy sensor to the PERFoRM Middleware. In particular, the implementation considers the requirements coming from the Roughness Process in the GKN industrial scenario (see Chapter 13). Here, a robotic cell with an industrial robot is used to execute processing and inspection operations on aeronautics parts. For inspection purposes, a Mitutoyo roughness sensor is mounted on the wrist of the robot.

In order to have the fully reconfigurability of the robotic cell, an important step is to substitute the cabled serial communication to the sensor with a wireless one. This can be easily achieved using a serial-WiFi converter, but in order to have completely no wires it must be battery-powered.

In Figure 5.5 an image of the hardware solution for the sensor Adapter is provided (from both conceptual and realization point of views).

From the software point of view, the sensor Adapter needs to implement an interface that permits to easily manage the sensor (configure, start the measurement, etc.) and collect the acquired data from external applications connected to the PERFoRM Middleware. For this reason, OPC-UA protocol

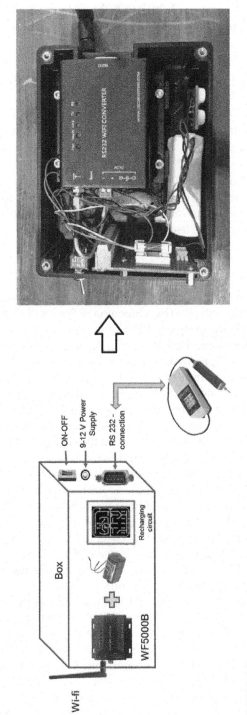

FIGURE 5.5

Sensor Adapter hardware for the GKN industrial scenario. (EU HORIZON 2020 FoF PERFoRM 2016.)

has been selected (Haskamp et al. 2017; OPC Foundation/OPC-UA 2018). The sensor Adapter implements an OPC-UA server which permits to

- start/stop the measurement;
- get/set measurement conditions (e.g., normative used, roughness profile, cut-off length, number of samples, etc.);
- get detector status: Idle (not moving in the home position), measuring (moving), retracting (going back to the home position); and
- get acquired data.

In Figure 5.6 the GUI of the sensor Adapter is presented.

Before the deployment of the sensor Adapter in the GKN robotic cell, the Adapter has been tested and validated in the MTC laboratory where communication and performance tests have been executed.

5.5.4 PLC Adapter

The PLC Adapter permits to connect any type of legacy equipment and newly built equipment controlled by a Siemens S7 Series PLC. In Europe, this is presently the most commonly used PLC in automation equipment such as assembly cells and lines, automatic quality control (QC) equipment, any type of dedicated process equipment and the like. Its market share in this specific segment of the automation industry is well above 50%.

Today's market requirements, with regard to data communication between the equipment level and the production department or factory level (the Enterprise Resource Planning [ERP] systems, the Manufacturing Execution System [MES] systems), are mostly focused on the equipment status information (machine ready—in production—not ready/error) and statistics concerning production lots (OEE level, machine availability, production data, etc.). With the arrival of Industry 4.0 the demand for data collection and communication has been rapidly increasing. Predominantly in the manufacturing of safety-related products such as air bags, breaking and locking systems in the automotive industry, or metered dose drug delivery systems in the medical field, parts traceability has become a frequently asked requirement. Such specifications require well-designed data structures and interfaces and are thus ideally suited as PERFoRM applications.

In the context of the applications of the PERFoRM approach it was initially foreseen to equip GKN's modular production and process cells with a Siemens S7 PLC to handle all auxiliary functions such as parts fixturing, controlling of peripheral equipment, monitoring of the safety function, etc. GKN, a manufacturer of components for the aeronautical industry is a typical representative of safety relevant products manufacturers (Company GKN 2018). The specific regulatory requirements in this industry call for stringent procedures, for instance, when a cell has to be reconfigured for the production of a

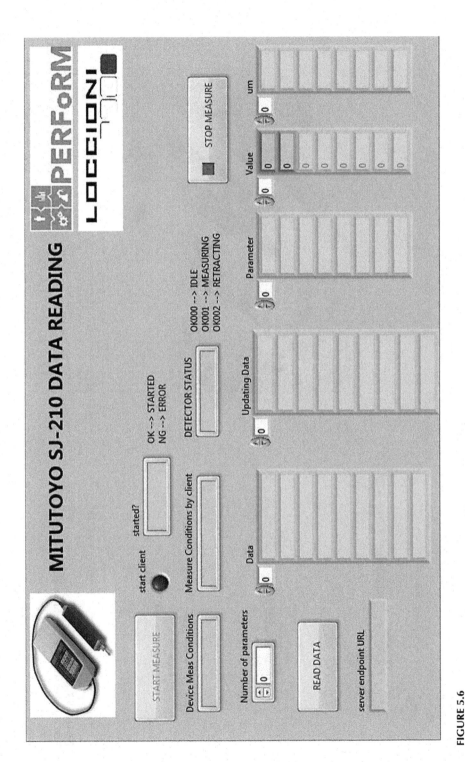

FIGURE 5.6
Sensor Adapter software Graphical User Interface (GUI) for the GKN industrial scenario. (From EU HORIZON 2020 FoF PERFoRM 2016.)

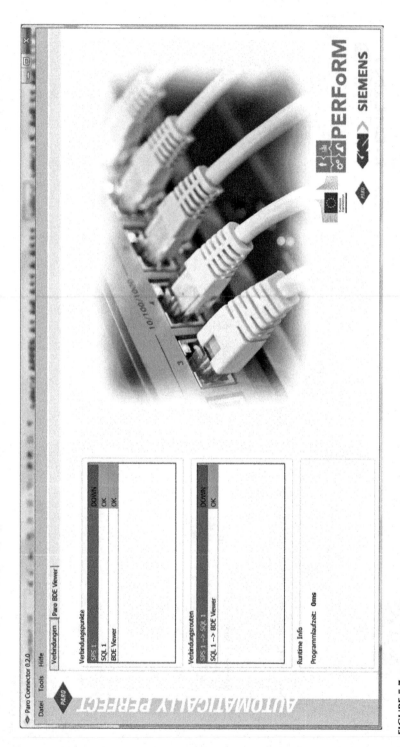

FIGURE 5.7
PLC Adapter software Graphical User Interface (GUI).

different component, all production recipes and setup parameters have to be downloaded from a safe, centralized program server. This procedure ensures that always the latest revision of the parts programs and parameters are used.

When GKN decided to switch to an ABB PLC, the development of the Siemens S7 PLC Adapter has been put on hold. PARO (PARO AG 2018), company responsible for the implementation of this Adapter, has instead focused its effort on the development of the MS SQL DB Adapter within the Siemens Compressors use case (Chapter 10). The development work accomplished up to this point will not be lost. PARO is striving to use it in the context of specific customer projects with high requirements for data exchange between the machine and the MES level.

The following description explains how the PLC Adapter can be configured and how it operates.

Physically the connection between the Siemens PLC and the Adapter can be either, for older generations of legacy systems, a ProfiBus wire connecting to the PLC communication module or, for newer systems (S7-1500 Series) an Ethernet wire connecting to the CPU module of the PLC.

The data communication is accomplished through socket connection defined by an XML-File between the Adapter and the PLC. Depending on the data structure available on the side of the legacy system, an adaptation of the PLC software is required.

The feasibility of this data communication concept has been developed and tried out in a test environment with a PC, serving as host for the PERFoRM Middleware component, and a PLC simulating the system's controller. It has been proven that the exchange of data functioned according to the performance criteria specified for typical real-life conditions.

The following figures show the GUI of the PLC Adapter (Figure 5.7) and its configuration (Figure 5.8).

FIGURE 5.8
PLC Adapter configuration: PC side (left) and PLC side (right).

5.6 Conclusion

The technology Adapters, described in this chapter, facilitate the integration into the PERFoRM platform of legacy hardware and software production systems: databases, robots, measurement sensors, and PLCs. The main role of these Adapters is to translate the device/module internal data model into the one defined according to the PERFoRM approach, but they can also provide additional features like the generation of a robot program from a 3-D model (Section 5.5.2) or the transformation of a wired communication into a wireless one (Section 5.5.3).

References

Angione, Giacomo, José Barbosa, Frederik Gosewehr, Paulo Leitão, Daniele Massa, João Matos, et al. 2017. "Integration and deployment of a distributed and pluggable industrial architecture for the PERFoRM project." *Procedia Manufacturing.* 11: 896–904. doi:10.1016/j.promfg.2017.07.193.

Chakravorti, N., E. Dimanidou, G. Angione, J. Wermann, and F. Gosewehr. 2017. "Validation of PERFoRM reference architecture demonstrating an automatic robot reconfiguration application." *IEEE Xplore.* 2017:1167–1172.

"Company GKN." 2018. Accessed October 10, 2018. https://www.gknaerospace.com/.

EU HORIZON 2020 FoF PERFoRM. 2016. "Deliverable D2.2: Definition of the System Architecture.". Unpublished manuscript, last modified October 10, 2018. http://www.horizon2020-perform.eu/index.php?action=documents.

EU HORIZON 2020 FoF PERFoRM. 2016. "Deliverable D3.1: Adapters implementation for at least three different production resources." Unpublished manuscript, last modified 10, 2018. http://www.horizon2020-perform.eu/index.php?action=documents.

EU HORIZON2020 FoF PERFoRM 2015-2018. Accessed October 11, 2018. https://cordis.europa.eu/project/rcn/198360_de.html.

Haskamp, Hermann, Michael Meyer, Romina Mollmann, Florian Orth, and Armando W. Colombo. 2017. "Benchmarking of existing OPC UA implementations for Industrie 4.0-Compliant digitalization solutions." Proc. of the 15th IEEE Int. Conf. on Industrial Informatics INDIN2017, pp. 589–594.

ISA Organization. 2018. "International Standard for the Integration of Enterprise and Control Systems: ISA'95." Accessed 10.October. https://www.isa.org/belgium/standards-publications/ISA95/.

MTC. U. K. 2018. "Manufacturing Technology Centre, UK." Accessed October 11, 2018. http://www.the-mtc.org/.

OPC Foundation/OPC-UA. 2018. "OPC-Unified Architecture." Accessed October 11, 2018. https://opcfoundation.org/about/opc-technologies/opc-ua/.

PARO AG. 2018. "PARO AG." Accessed October 11, 2018. https://www.paro.ch/.

6

Architectural Components: Middleware

**Jeffrey Wermann, Frederik Gosewehr, Waldemar Borsych,
Armando W. Colombo**
(Institute for Informatics, Automation and Robotics (I²AR))

CONTENTS

6.1 Introduction

One of the core components of the architecture envisioned for the Production harmonizEd Reconfiguration of Flexible Robots and Machinery (PERFoRM) approach is a common integration platform. This platform, called the Middleware, must ensure that each PERFoRM-compliant system, tool, or device, located on the OT (Operational Technology/shop floor) level or on the IT (Information Technology) level can communicate with each other, guaranteeing both "Connectivity and Interoperability" (Givehchi et al. 2017). While within PERFoRM's architecture design the way of how data is structured is defined by a generic data model and a set of necessary interfaces is given, the exact implementation of the communication (e.g., which protocols are used) is deliberately not fixed. This allows more flexibility for the developer of a PERFoRM-compliant solution, giving freedom related to which component or systems can be integrated. The role of the Middleware is to translate between different incoming and outgoing communication technologies, acting as a message broker. Additionally, the Middleware should have capabilities to not just be a simple protocol translator, but also to offer more sophisticated routing features, allowing complex message flows and data manipulation while processing incoming data (EU HORIZON 2020 FoF PERFoRM 2017).

The next sections describe an approach to adapt existing Middleware solutions to comply with PERFoRM's architecture and support PERFoRM's standard interfaces. Additional to the analysis of existing commercial solutions, one specific Middleware implementation is described, to be used as a proof of concept and a basis solution for an application in the various test beds and the use cases presented in Chapters 10–13 of this book.

6.2 Role of the Middleware

In Computer Science, the term "Middleware" describes a software layer or component, which allows different software applications to interact with each other (Etzkorn 2017). Typically, large systems consist of many different software applications, which—in most cases—do not provide specific interfaces to each other but implement individual interfaces. A Middleware is used in these cases to make these interfaces fit together. This can be useful as an additional component on top of an operating system, which will link different applications in one computer together, but is also very useful in networked systems, where the applications are running on separate hardware. The advantage for the software developers is that they no longer need to implement specific interfaces for each software it needs to interact with; only one interface is required, which the Middleware can handle.

The architecture of PERFoRM is illustrated in Figure 6.1. It represents the different components that are part of a PERFoRM-compliant production system and how they are linked with each other. The different components foreseen to be a part of the architecture are applications working on different layers of the ISA'95 reference architecture (Enterprise-control system integration: ANSI/ISA-95.00.01-2000 2000). On the production level, different production components, such as Programmable Logic Controllers (PLCs),

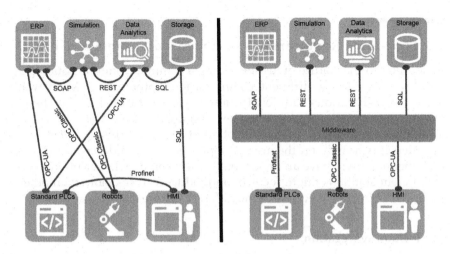

FIGURE 6.1
A system without Middleware (left) and with Middleware as an integration layer (right).

Computerized Numerical Control (CNC) machines, and industrial robots, must be supported. This also includes the integration of smart production components, for example, based on agent-based approaches. On the IT level, various management tools, such as Enterprise Resource Planning (ERP) and the Manufacturing Execution System (MES) systems, are integrated. Furthermore, the various tools developed within PERFoRM, which target simulation, data analytics, and data visualization, are an important part of the architecture.

The PERFoRM architecture is following a Service-oriented approach, which means that each component within the architecture is exposing its functionality as a service, using a specific data model (Peres et al. 2016). Newly developed tools will follow this principle by design. But since legacy systems, especially on the production level, are not supporting these Service-oriented technologies, additional adapters need to be developed to enable these components to follow the principles of a Service-oriented architecture (SOA).

A core component of the architecture is the Middleware, which acts as an integration platform for the individual components. By using a Middleware, a loose coupling of each individual component can be achieved. Instead of programming multiple interfaces per component that connect to the multiple other components it needs to communicate with, only one interface to the Middleware needs to be developed. The Middleware can be configured to route the messages to the right destination and to translate the message to the right protocol, which the destination system is able to understand.

6.3 Functional Requirements for the Middleware Specification

In addition to the feature of just providing a common communication platform and translating different communication protocols, a Middleware can come with various additional features (Gosewehr et al. 2017). The following chapters will give an overview about the different aspects, which a Middleware solution can cover. Not all of these features are mandatory, as it depends heavily on the area of application of a Middleware. This on the other hand will have an impact on the decision on what solution to use, as different solutions might focus on individual features more than others. Figure 6.2 shows an overview of the different feature clusters.

6.3.1 Data Aggregation

One of the main features of a Middleware is that it can gather data from various data sources. These sources can be software applications or hardware devices, which are enhanced with software adapters for communication. Furthermore, it is also necessary that the Middleware is able to communicate with other data integration technologies, such as other Middleware solutions, communication architectures like Open Platform Communications Unified Architecture (OPC-UA) (Damm, Leitner, and Mahnke 2009) or through message broker technologies, such as Message Queuing Telemetry Transport (MQTT) or Advanced Message Queuing Protocol (AMQP).

The Middleware needs to be able to send and receive data from these components, using different methods, such as polling or subscription. Polling describes the periodical request of data, whereas the subscription concept is

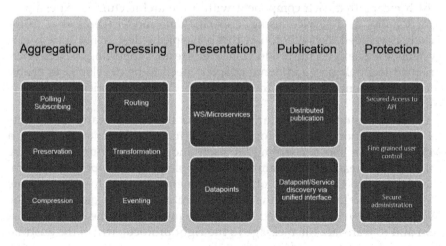

FIGURE 6.2
Middleware features.

implementing a way to subscribe only to changes of specific data, reducing the network traffic.

Furthermore, a Middleware is also able to preserve the received data by storing it temporarily within a history buffer during the transmission of the data. This is used to buffer the data to be retrieved in case of loss of data or to increase the data reliability in general. As an additional step, the received data can also be compressed by only storing data that is important, for example, the data that has changed its value. The rules for the compressing and storing of data need to be configured within the Middleware.

6.3.2 Data Processing

After receiving data from an application, this data needs to be processed and forwarded by the Middleware. In this phase, the Middleware is acting as a router, which defines which incoming data will be forwarded to which target application. Multiple targets are also possible. In some cases, Middleware solutions provide graphical interfaces to draw these routes, whereas in other cases the routes need to be configured otherwise (e.g., configuration files). The Middleware is not able to decide the routes by itself. Therefore, if a new component enters the system and needs to send data via the Middleware, this needs to be configured within the Middleware.

Within this step, it is also possible to manipulate the data itself. This can be used to transform data from one data format to another, for example, in cases where the incoming data is formed in a way the sending application is understanding it (e.g., XML), whereas the receiving application is expecting the data to be in a different format (e.g., JSON). This transformation also needs to be configured, using languages such as XSL Transformation.

A Middleware can also implement an event-based communication. For example, how the Middleware can aggregate data (see Section 6.3.1) and support the subscription feature also for the communication partners. Instead of always directly requesting information, a connected client application can now subscribe to specific events. These events can be status notifications or changes of important values. Should a status or value change occur, all subscribed clients will be notified accordingly. The Middleware needs to be able to implement this feature for applications that do not provide it by themselves, and it needs to provide a common interface for all applications to use this subscription model.

6.3.3 Data Presentation

The Middleware must be able to present data in different ways. Two major ways of presentation should be implemented to allow an easy integration:

- Web services
- Data points

Web services are functions that are provided by a service provider (e.g., a machine or a software application) and are accessible through the network, allowing a machine-to-machine communication. Web services are typically described in XML or JSON, providing an object-oriented interface. They can use various communication protocols, including XML-RPC, Simple Object Access Protocol (SOAP), and Representational State Transfer (REST). Using Web services is commonly used in informatized systems, especially from the business layer, to achieve a loose coupled communication.

Data points are a more static way to provide data. A server is configured to publish a predefined set of data, which can be accessed using a specific protocol, which can vary heavily depending on the application or machine that is providing the data points. Examples for this can be PLCs or supervisory control and data acquisition (SCADA) systems, which can publish specific process data or databases, which can offer data points through an SQL interface. This way of accessing data is very common in automation devices and other low-level industrial components.

As both approaches are greatly accepted and common in different layers (production and business) of an industrial system, the Middleware is playing an important role by translating between these two presentations.

6.3.4 Data Publication

A Middleware is targeting a loose coupled approach, where individual applications don't necessarily need to know details about each other. It additionally can provide services to publish and discover connected services. This means that each component, which acts as a server and provides a service (either a Web service or a simple data point), can register the said service within the Middleware. Other components are now able to discover this service by using discovery mechanisms. This way, applications can interact with each other dynamically and find each other during runtime without having to preconfigure to which exact communication partner it must connect to.

To implement such a system, different approaches are possible. The first one is to publish new services automatically to all connected applications, so that each application will be notified about each new service within the system and react accordingly. A second solution is to register the services in a DB within the Middleware itself. Applications can now proactively request which services are available in the system. This mechanism is often compared to the "Yellow Pages" for telephone numbers and addresses and will therefore be referenced as such throughout this document.

6.3.5 Data Protection

An important issue of all networked systems is the security. The Middleware must provide mechanisms to ensure a secure communication between all connected components. This includes the transmission security, which can be achieved by using various data encryption and other secure transmission methods. Furthermore, it is necessary to ensure a controlled access to data by being able to regulate which application can use a specific service or access a specific data point. Additionally, the Middleware itself must be secure, so that only restricted personnel can (remotely) reconfigure the Middleware.

While the topic of data protection is an important part of a Middleware and will play a role in the evaluation of possible Middleware solutions, this project will not discuss the topic in depth. The focus will be on using technologies provided in the existing solutions and not on developing new security mechanisms.

6.4 Middleware Design

A first investigation of the existing Middleware solutions has shown that the market already offers a big set of solutions to be used as Middleware or Enterprise Service Bus (ESB). The investigation has also shown that these solutions can vary a lot in the functionalities they provide and the application areas they are targeted at. In some cases, there might be a need for a solution to integrate multiple shop floor devices distributed throughout a factory, whereas in other cases the connection between different IT systems is necessary. It is impossible to locate a software, which is the perfect solution to be used in every application area. Furthermore, from a more practical point of view, it is impossible to pick one specific solution and force all the use case providers to use it, since all of the solutions detected are coming with big investments both in the acquisition of the software itself and especially in the implementation of the said solutions into their existing production systems. In some cases, Middleware solutions might even already exist.

Therefore, the approach for the Middleware within PERFoRM was to make the exact solution used in each specific application flexible and to provide a way to integrate these existing solutions in a way that they are harmonized with the requirements of the PERFoRM's architecture. Figure 6.3 shows the basic design of the Middleware envisioned for PERFoRM. The different components will be described in the following chapters.

6.4.1 Shell Middleware

Since the individual core solutions can vary a lot in the functionalities they provide from the start and which communication protocols they support, an additional integration layer has been added for the PERFoRM Middleware. The so-called Middleware shell is a set of software tools that will be used to harmonize the individual core solutions. This way, it will be ensured that independent of the core solution selected, each Middleware will act the same from the point of view of the components outside of the Middleware. This shell will

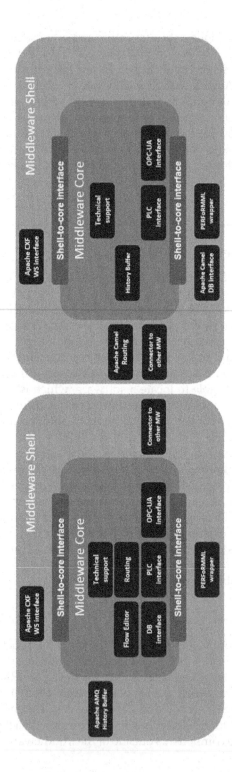

FIGURE 6.3
Inner architecture of the Middleware (two possible examples based on a selection of functionalities).

fill the gaps that the selected core solutions might provide, by implementing missing functionalities and adding new communication technologies.

The shell is built in a modular way and already can provide all the functionalities required for the PERFoRM Middleware. It is missing the commercial support and easy integration tools with graphical user interfaces, which are provided by the core solutions. Therefore, the use of just the shell for a full Middleware implementation is possible and effective for testing the integration of specific tools with the Middleware but is not advisable for a real industrial application.

The implementation of this shell has been done using Apache ServiceMix, which is described in Section 6.5.

6.4.2 Core Middleware Solutions

Since PERFoRM is an innovation project (EU HORIZON 2020 FoF PERFoRM) (HORIZON2020 FoF Project 2015–2018), and is therefore searching for solutions that can be directly applied within real industrial applications, the focus of the search for the right Middleware for being used in the PERFoRM approach lies on already existing and established Middleware solutions. Developing a new Middleware from scratch is out of scope of the PERFoRM approach and would not lead to the necessary maturity to make it applicable within the four industrial use cases.

Therefore, an analysis of existing Middleware solutions has been carried out (Gosewehr, Wermann, and Colombo 2016). The goal was to select one solution to fit all the possible requirements. If there is no software which fits for all use cases as Middleware, it is necessary to at least select one solution per application area. These will then act as a possible core of the Middleware design.

Since the market is full of software programs that claim to be Middleware solutions to different extents, the evaluation of solutions was carried out in two steps. In the first step, a pre-assessment of all major Middleware software solutions was done, which consisted of a superficial analysis to reduce the amount of solutions to a manageable amount. This was followed by an in-depth analysis, where the most promising solutions were installed and tested by different partners within the project. Based on these tests, a final evaluation of the Middleware solutions has been done.

6.5 Apache ServiceMix

Apache ServiceMix is an open-source container including various solutions from the Apache software family to build a SOA-based ESB. It is lightweight and easily embeddable. It supports the Open Services Gateway Initiative (OSGi) framework to allow modularity. The ServiceMix itself provides several

Java-based frameworks and tools to implement message broker functionality (Apache ActiveMQ), routing and integration patterns (Apache Camel), and Web service-based interfaces (Apache CXF). It is highly adaptable, but needs a considerable amount of expertise to both extend and configure the ESB. Graphical configuration tools (e.g., flow editor for route configuration) are not included. Configuration is usually done using XML descriptors with the Spring and XBean frameworks. The ServiceMix doesn't include interfaces to production devices, but new interfaces can be implemented in Java with existing, tested protocol stacks and easily integrated in the system.

6.6 Middleware Implementation

The actual Middleware, as previously explained, is based on an outer shell, the so-called integration layer and, if needed, the internal "core" Middleware, which can be any kind of system able to process incoming and outgoing information "in between" (Gosewehr et al. 2018). To achieve such functionality, the integration layer, which to the outside world is only a well-defined Application Programming Interface (API), has been developed using the Apache ServiceMix ESB. This implementation, which is described on the next, showcases a methodology to integrate consumers and providers but is not limited to ServiceMix. It could also be implemented in other Middleware tools such as Microsoft Biztalk, IBM Integration Bus, or even WinCC OA.

A direct integration of the resulting code should also be possible if the Middleware utilizes OSGi, as the PERFoRM Middleware forms an OSGi bundle (a bundle of Java archives that form an executable unit). This is especially possible when the Middleware uses Apache Felix, which is the OSGi implementation Apache ServiceMix and underneath it Apache Karaf as MicroKernel. Examples of these Middleware systems are Tibco, WSO2, or mBS from Prosyst (subsidiary of Bosch).

The *Core Middleware* implementation, be it IBM IB, WinCC OA, or other "Middleware" solutions, can intercept incoming and outgoing messages to further process data or enrich it with additional content. To do so, the core needs to register an interception route within the integration layer to declare an additional step within the routing slip (for Hypertext Transfer Protocol (HTTP)-based services). For the connection to the so-called "data points," for example, to sensory data, a message broker is used, which connects an HTTP client to the specific data point. Such data points are first registered within the integration layer via Application Modelling Language (AML) and may be queried from within the core to create a new route. Such a rerouting would force the integration layer to query the data point's information from the core instead of the actual provider. The actual rerouting is not automatically configured. New routes need to be added manually after changes to the data model have been made. This can happen during runtime using the Aries Blueprint XML descriptions within the ServiceMix's deploy folder or statically via additional Java bundles. Although an automatic route generation would be preferable, it has been stripped from the design as this point depends heavily on the specific Middleware implementation and might not be replicable on other mediation engines not based on Apache Camel. ple Talend ESB, WSO2, or Red Hat JBOSS Fuse.

6.7 Conclusion

This chapter has described how a Middleware should be used within PERFoRM as an integration platform to enable the communication, connectivity, and interoperability between the various components foreseen within the PERFoRM architecture. To achieve a common understanding of what the Middleware is and what it can do, the basic functionalities of the Middleware have initially been defined and described.

These functionalities were the requirements for evaluating possible solutions to be used within the approach. As an innovation approach with strong connection to prototype implementations in real industrial environments, an additional requirement of major importance has been the applicability in real industrial manufacturing systems. This, has limited the selection to solutions for which the distributors are offering industrial support and which are already proven in productive use. Based on these requirements, a set of possible solutions has been selected and described in this chapter. These solutions meet the requirements to an acceptable extent and are targeting different application areas. All of them can be configured in a way that they comply with the PERFoRM architecture, which means that they support a Service-oriented approach and can handle messages following the standard interfaces and data models specified and described in Chapters 3 and 4 of this book. These solutions are just a set of possible implementations, but other implementations are not excluded from being used, as long as the compliance to the PERFoRM architecture described in Chapter 3 can be ensured.

Additional to the selection of various possible solutions, one instance of the Middleware based on Apache ServiceMix has been implemented. The Apache ServiceMix solution acts as a reference implementation, which will be used during the project as a demonstration and test environment. It lacks the industrial support, but provides all the functions that are envisioned for a PERFoRM-compliant Middleware component. It is open to add more functionality and provides its key features in a modular way. For this reason, another important application of ServiceMix is to bridge any gaps that the existing Middleware solutions provide. This chapter provided an overview on how the Apache ServiceMix-based Middleware works, how it is configured, and how external components can interface with it.

In summary, a PERFoRM Middleware specification has been defined, and multiple solutions that are capable of fulfilling these requirements have been evaluated. An Apache ServiceMix-based example implementation has been set up, which fulfills all the requirements and is configured in a way to suit the PERFoRM architecture.

References

Damm, Matthias, Stefan-Helmut Leitner, and Wolfgang Mahnke. 2009. *OPC Unified Architecture*. Berlin, Heidelberg: Springer-Verlag Berlin Heidelberg. http://dx.doi.org/10.1007/978-3-540-68899-0.

Enterprise-control system integration: ANSI/ISA-95.00.01-2000. 2000. Research Triangle Park, N.C. Isa.

Etzkorn, Letha H. 2017. *Introduction to Middleware: Web Services, Object Components, and Cloud Computing*. Boca Raton, FL: Chapman and Hall/CRC.

EU HORIZON 2020 FoF PERFoRM. 2017. "Deliverable D2.4: Industrial Manufacturing Middleware: Specification, Prototype Implementation and Validation.". Unpublished manuscript, last modified October 10, 2018. http://www.horizon2020-perform.eu/index.php?action=documents.

"EU HORIZON 2020 FoF PERFoRM." Accessed October 10, 2018. https://cordis.europa.eu/project/rcn/198360_de.html.

Givehchi, Omid, Klaus Landsdorf, Pieter Simoens, and Armando W. Colombo. 2017. "Interoperability for industrial cyber-physical systems: An approach for legacy systems." *IEEE Trans. Ind. Inf.* 13 (6): 3370–3378. doi:10.1109/TII.2017.2740434.

Gosewehr, Frederik, Jeffrey Wermann, and Armando W. Colombo. 2016. "Assessment of Industrial Middleware Technologies for the PERFoRM Project." In *Proceedings of the IECON2016—42nd Annual Conference of the IEEE Industrial Electronics Society: Florence (Italy), October 24–27, 2016*. pp. 5699–5704. Piscataway, NJ: IEEE.

Gosewehr, Frederik, Jeffrey Wermann, Waldemar Borsych, and Armando W. Colombo. 2017. "Specification and Design of an Industrial Manufacturing Middleware." In *2017 IEEE 15th International Conference on Industrial Informatics (INDIN): University of Applied Science Emden/Leer, Emden, Germany, July 24–26, 2017: Proceedings*. pp. 1160–1166. Piscataway, NJ: IEEE.

Gosewehr, Frederik, Jeffrey Wermann, Waldemar Borsych, and Armando W. Colombo. 2018. "Apache Camel Based Implementation of an Industrial Middleware Solution." In *Proceedings 2018 IEEE Industrial Cyber-Physical Systems (ICPS): ITMO University, Saint Petersburg, Saint Petersburg, Russia, 15–18 May 2018*, pp. 523–528. Piscataway, NJ: IEEE.

HORIZON2020 FoF Project. 2018. "PERFoRM." Accessed October 10 2018. http://www.horizon2020-perform.eu/.

Peres, Ricardo S., Mafalda Parreira-Rocha, Andre D. Rocha, José Barbosa, Paulo Leitão, and José Barata. 2016. "Selection of a Data Exchange Format for Industry 4.0 Manufacturing Systems." In *Proceedings of the IECON2016—42nd Annual Conference of the IEEE Industrial Electronics Society: Florence (Italy), October 24–27, 2016*. pp. 5723–5728. Piscataway, NJ: IEEE.

7

PERFoRM Methods and Tools

Sebastian Thiede, Lennart Büth, Benjamin Neef
(Technische Universität Braunschweig)

Phil Ogun, Niels Lohse
(Loughborough University)

Birgit Obst
(Siemens AG)

Mostafizur Rahman
(The Manufacturing Technology Centre (MTC))

José Barbosa
(Instituto Politécnico de Bragança)

CONTENTS

7.1 Introduction

The Production harmonizEd Reconfiguration of Flexible Robots and Machinery (PERFoRM) architecture is based on bringing diverse (machine or other) data streams into a consistent Middleware component. From that, a variety of different applications can be developed that take the data needed in order to support the planning and operation of production systems, from the perspective of the Life Cycle and Value Stream specifications of the cyber-physical systems (DIN/VDE 2017; IEC 2017; McKinsey & Company 2015) (see Figure 5.1 in Chapter 5).

Within the PERFoRM project, different digital methods and tools for the support of production planning have been developed and implemented. Examples are tools to automatically compile simulation models from the Middleware component to specific data analytic methods for predictive maintenance forecasting, and to the visualization of plant-relevant data for supporting the reconfiguration of production systems.

For an initial overview, the developed methods and tools and their related functionalities can be allocated within an effort/complexity matrix adapted from Barga et al. (2014) (Figure 7.1). Benefit is related to the potential contribution toward a flexible and reconfigurable manufacturing system, whereas effort is related to the needed data and complexity of the solution. Different evolutionary steps can be distinguished in the illustration. It starts with providing Key Performance Indicators (KPIs) that give transparency of the current situation (*descriptive*). While this step just describes the as-is situation, the next evolutionary step gives further *diagnostic* information on the root causes; for example, based on pattern recognition. Both steps are rather retrospective and do not provide prognosis functionalities. Therefore, *predictive* components, such as simulation, need to be integrated that allow the evaluation of alternative future scenarios. In the last step, the integration of knowledge bases (e.g., on possible improvement measures) allows active decision support in the sense

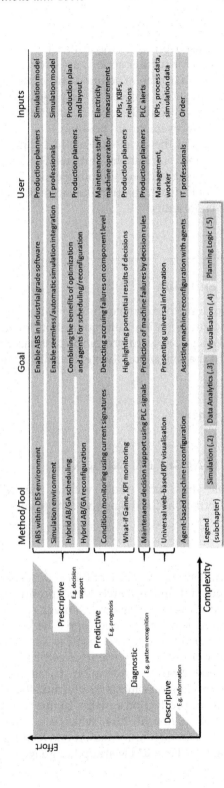

FIGURE 7.1
An overview of selected solutions developed within PERFoRM classified on the basis of potential effort/complexity, left side. (From Barga et al. 2014.)

of *prescriptive* analysis—based on as-is and estimated future situations, favorable recommendations for further actions are provided.

Within the matrix (Figure 7.1) different solutions from the PERFoRM project are classified—while involving a diversity of project partners, quite a range of functionalities are covered. Therewith, applications for different stakeholders and areas of improvement are possible.

In this chapter, a selection of the developed methods and tools is briefly presented and categorized into simulation solutions (Section 7.2), data analytic solutions (Section 7.3), visualization solutions (Section 7.4), and planning logic solutions (Section 7.5). For each solution the target definition is formulated before the solution is presented.

7.2 Material Flow Simulation

The PERFoRM approach aims to automate the generation of material flow simulations within the Information-Communication-Technology (ICT)-Architecture and to make a step toward bringing agent-based methods and tools tailored for industrial applications. Two exemplary solutions are presented in this section.

7.2.1 Simulation Environment

TARGET DEFINITION

Seamlessly integrates simulation tools within the Middleware. Enabling on-the-fly simulation runs using live data from the manufacturing system.

Principally, a generic Simulation Environment (SE) is developed that allows the integration of commercial and research-based production simulation tools in a service-oriented manufacturing environment, such as in Industry 4.0 factories, to support flexible reconfiguration and scheduling decisions on the industrial ICT platform. For this purpose, the SE is connected to a generic Middleware component and processes parameterized simulation requests via automated model generation and execution, based on current shop floor conditions. These simulation requests are formulated in a standardized description language for production systems, in this case a derivative of AutomationML (AML) (AutomationML 2014), which was extended by simulation-specific data sets, to exchange information between the Middleware and the services utilizing the PERFoRM Markup Language (PML).

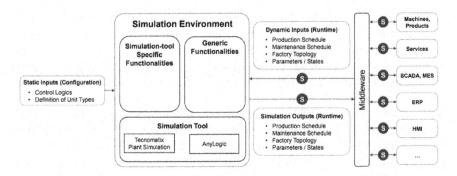

FIGURE 7.2
Context of simulation environment, implementation view. (From Fischer, Obst, and Lee 2017.)

Figure 7.2 provides an informational view of the proposed architecture context for the SE, illustrating the separation of the main functionalities of the SE into submodules as well as the flow of various information types into and out of the SE. The SE contains a discrete-event-based, logistics Simulation Tool such as Tecnomatix Plant Simulation (Siemens 2018) or AnyLogic (Anylogic 2018). The SE wrapper architecture captures as much functionality as possible into a set of tool-independent Generic Functionalities regarding interfaces, coordination, pre- and post-processing services, and library management, which would support a multitude of various simulation tools in an independent manner. Additionally, a set of simulation tool Specific Functionalities are also included, mainly to conduct actual simulation runs within the provided simulation engine (Fischer et al. 2017). With the simulation service request via the Middleware component, the required simulation model of the production system is updated or generated automatically, using configuration inputs, current production/maintenance schedules, and actual system state for initialization. After execution, the schedule KPI results are then sent back to the Middleware (via the standard interface) to support optimized scheduling and reconfiguration issues on the fly.

7.2.2 Simulating Agent-Based Systems within Discrete-Event Software

TARGET DEFINITION

Development and testing of a method to use the agent-based simulation (ABS) paradigm within industrial grade discrete-event simulation (DES) tools.

ABS has been adopted by simulation practitioners to simulate complex coherences in manufacturing systems. Especially with the trend of flexible manufacturing, promoted through Industry 4.0, ABS can support the evaluation of complex situation, when planning and/or scheduling production within flexible manufacturing systems. To apply ABS in manufacturing, supporting tools are necessary. At present most manufacturing system-specific simulation tools are based on DES. These industrial simulation tools offer a hands-on user interface (UI), and thus enable the industry to model situation in an efficient way. Tools that are normally used to do ABS are more general-purpose simulation tools and are mostly used in research due to their complex application characteristics. Thus, in order for the industry to be able to use ABS and its advantages, they must be able to apply ABS within commonly used industrial simulations tools.

Therefore, a method was developed and tested to utilize the ABS paradigm within a DES environment. Figure 7.3 (on the left) shows the three main steps to transform the elements of a DES into an ABS logic. Furthermore, the left-hand side of the figure exemplifies the general elements of a DES, and the right-hand side displays the functions of the simulation software that are used. The breakdown into these three steps results from the simple model of an agent and its interactions. In the first step, the environment is built, in the second step the agent is implemented, and in the third step the interaction between the agent and environment is introduced. This procedure was tested in a case study, migrating an ABS from the software tool AnyLogic (Anylogic 2018) to the DES simulation tool Tecnomatix Plant Simulation (Siemens 2018); the specific procedure is highlighted in Figure 7.3 (right) (Büth et al. 2017).

7.3 Data Analytics

The PERFoRM approach aims to utilize the power of an integrated Middleware component by using connected data analytic approaches.

7.3.1 Computer Nummerical Control (CNC) Machine Condition Monitoring with Current Signatures

TARGET DEFINITION

Component-based condition monitoring of CNC machine tools by the use of main terminal high-frequency current signatures.

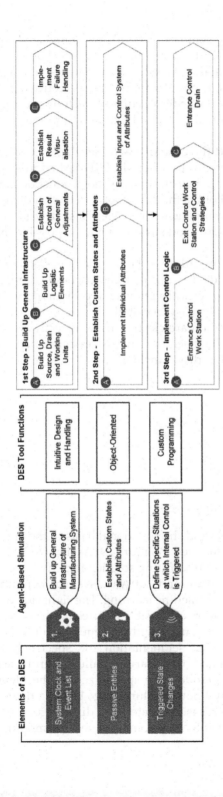

FIGURE 7.3

Three general steps for implementing an agent-based logic into a discrete-event simulation tool (left); Case study: Conceptual roadmap for the implementation of an agent-based matrix manufacturing system into a discrete-event simulation tool (Tecnomatix Plant Simulation) (right). (From Büth et al. 2017.)

Production facilities are complex and highly automated combinations between mechanical and electrical components. A breakdown results in significant economic losses, material losses, environmental damage, and in the worst case to bodily harm. Especially from the perspective of the personnel, less production lines, the knowledge of current machine and component status, and the knowledge about the actual tool wear and the product quality are indispensable.

Therefore, the topic and the objective is to develop a condition monitoring system in accordance with the overall PERFoRM approach, in order to detect anomalies regarding the condition of the production equipment. By using the electrical signal of the machine tool, a detection of anomalies and defects down to single components (e.g., electrical drives, pumps, linear drives) and of tool wear status and surface quality is possible (Neef, Bartels, and Thiede 2018).

A big advantage of this procedure is the simplicity in electrical data acquisition in contrast to other data acquisition approaches (e.g., vibration monitoring). Measuring of the electrical signal is not position-dependent and the hardware is unexpansive and easy to install.

In focus of the investigation are electric drives directly connected to the electrical distribution of the machines and drives connected via frequency converters. Typically electric drives include

- Induction motors,
- servo motors,
- linear drives,
- pump motors,
- hydraulic pumps,
- lubrication pumps, and
- cooling lubricant pumps.

The basic application cycle and the functional basis of the approach in the context of PERFoRM are depicted in Figure 7.4. As depicted on the left side, the solution is designed as a retrofitting solution for existing CNC machine tools and consistent hardware and software system including measurement and analysis based on machine learning. The right side represents the toolchain for the identification of possible failures on the component level. Dedicated test runs are measured with a frequency of 10 kHz (current and voltage per phase). After segmentation and spectrogram extraction the data is fed to a pretrained deep learning failure identification module. Pretrained deep learning models are specialized in the identification of failures regarding already known failures (e.g., bearing defects). By feedforward a set of raw features is calculated and directly forwarded to a rest interface and as well fed into an anomaly detection module to identify

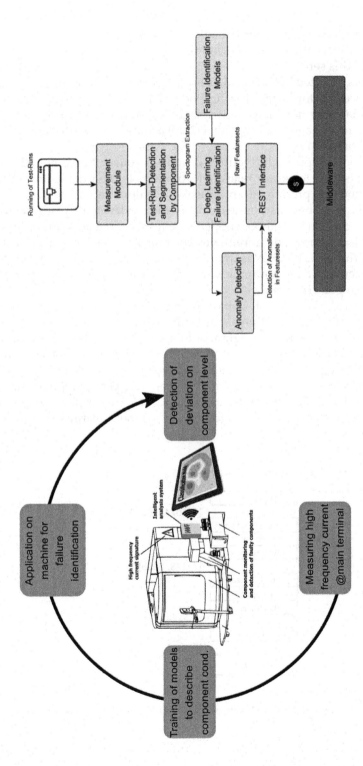

FIGURE 7.4
Application cycle of the proposed approach (left); toolchain for failure identification from pretrained model in context of the PERFoRM approach (right). (From EU HORIZON 2020 FoF PERFoRM 2017).

deviations from working order behavior. The deep learning failure identification module with pretrained failure identification models is vitally important in this setup.

Approaches reported by Shin et al. (2016) and Yanai and Kawano (2015) use a similar procedure for image classification tasks. Classification accuracy for unknown images or non-congeneric images increases significantly. By the use of pretrained ImageNet data, Yanai (2015) achieved a classification accuracy of 99% for food images.

In the proposed approach (Figure 7.4, right), a convolutional neural network (CNN) is used to extract relevant feature sets for failure identification. The CNN model is pretrained on an isolated failure pattern to achieve a discriminability as best as possible and is then evaluated by segmented spectrogram data from component-dependent measurements. Pretraining of the CNN is conducted by minimizing the binary cross-entropy loss with an Adam optimizer. The CNN is trained for a maximum of 45 epochs. If the validation accuracy does not improve for at least five epochs, the training stopped as a regularization method.

The pretrained CNN operates on overlapping red, blue, and green (RGB) image slices with 1024×100 pixels in feedforward. The RGB image slices were obtained by an RGB spectrogram of 1024×4120. The cuts are performed with an overlap of three discrete timestamps resulting then in 1341 segments of 1024×100. The proposed CNN has two convolutional layers with a hyperbolic tangent as the activation function: A dropout layer for regularization to avoid overfitting and a batch normalization and flattening layer to reduce the feature dimensionality to an output of 256 synthetic features. Figure 7.5

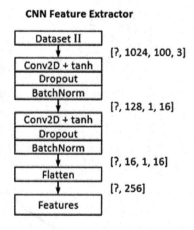

FIGURE 7.5
Architecture of Convolutional Neural Network (CNN) classifier for feature extraction (own illustration).

represents the architecture and the feature dimensions used in the feedforward CNN for feature extraction.

7.3.2 Maintenance Decision Support Using PLC Signals

TARGET DEFINITION

Alarms and failure analysis for the purpose of maintenance decision support.

The manufacturing industry is facing myriad issues regarding the availability and reliability of its equipment. Equipment faults can cause long production line stoppages, high maintenance costs, and low product quality. Well-planned maintenance assists in keeping the equipment in a healthy condition, and helps to decrease the risk of large-scale machine damages and revenue loss.

Condition-based predictive maintenance is considered to be the best approach for improving equipment reliability, reducing production costs due to failure, reducing costs due to maintenance, optimizing maintenance intervals, reducing the risks of catastrophic damage of the health of the machines, and minimizing unplanned downtime. However, in real production environment not all the failures and downtime reported are due to the condition of equipment. Many of the failures and downtime are caused by the production process itself rather than the condition of the machines. Hence, proper insight into the problem by analyzing historical data is very important for selecting an appropriate maintenance task.

The Maintenance Decision Support and analytics tool has been developed to explore the historical data related to failures and machine conditions (alarms) to get proper insight into the problem for guiding the future maintenance tasks. Data mining is defined as the exploration and analysis, by automatic or semiautomatic means, of large quantities of data stored in databases. Its main focus is the discovery of useful knowledge, including meaningful patterns and rules, from raw and apparently unrelated data. The proposed framework also utilizes data from appropriate maintenance manuals. The Data Analytics tool is responsible for (1) preprocessing the data from the failure records and maintenance tickets and alarms, (2) aggregating the data, and (3) identifying trends within the data. Details of each block can be seen in Figure 7.6(Chakravorti et al. 2018).

Failure grouping: The Maintenance database contains description of the problem reported in an unstructured textual form. Though this textual form is not the failure type, it may give an indication of the failure type. Consequently, it was necessary to label the information in order to determine

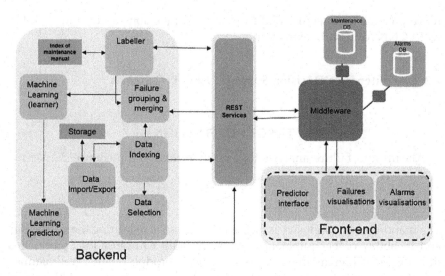

FIGURE 7.6
Toolchain for maintenance decision support using PLC signals. (From Chakravorti et al. 2018.)

the type of the machine failure from the textual form (the reported problem). Information of the prescribed machine failures reported in the Maintenance manual is being used for labeling the failure records; therefore, the inputs for the Backend System are the data from the Process Data Acquisition (PDA)/ Machine Data Acquisition (MDA), Manufacturing Messaging Specification (MMS) systems and the Maintenance manual.

Data indexing: Millions of alarms have been recorded in the database systems. An efficient way of querying the database was needed to handle the high-volume data in a very short time. Once the data are grouped and merged, an indexing operation is used to process the data using dates. The purpose of indexing is to enable a faster response to queries. Consequently, the selection of the relevant data (depending on the visualizations) can be done in a few memory accesses. In-memory indexing allowed an increase in speed of a magnitude of 10–100 times compared to standard database accesses, thus allowing real-time visualizations.

Front end: This module is used for visualizing the failures and the alarms dataset. The visualization of failures and alarms will provide a useful insight on the past events and also support the generation of appropriate maintenance tasks. The Front end uses the preprocessed data and creates visualizations for the failures and the key alarms. The purpose of this visualization is to monitor the condition of the machine by analyzing the trends of key alarms and potentially predict potential failures.

Machine learning for predictive model: After preprocessing (labeling and combining the alarms and failure data) and balancing the data, machine learning techniques are applied to generate a predictive model to extract decision rules. The Decision trees models are commonly used in data mining to explore and classify the data. These decision rules can also be used for conducting root cause analysis of the failures and also for the implementation of a decision support system that enables the detection of failures. Identification of key alarms associated to a particular failure can also be conducted.

7.4 Visualization

One of the key features of the PERFoRM architecture is the unified and easy to access data. To take advantage of these data streams different visualization tools were developed. Two examples are presented in the following subsections.

7.4.1 KPI Monitoring with What-if Game Functionality

TARGET DEFINITION

Dynamic monitoring of KPIs and Key Business Factors (KBFs) for tactical and strategical decision support.

The KPI monitoring with the what-if-game functionality tool (KPI tool or What-if tool, for short) is a Web-based solution to support decision-making strategies. Defined KPIs are continuously monitored, enabling the detection of trends and deviations. On this basis, the tool also enables the user to perform what-if games based on the variation of KBFs and generating the associated KPI implication. Therefore, the decision makers are able to verify the impact of changing the KBFs at the KPI level in an intuitive manner. Architecturally, the tool is composed by the KPI calculation and statistical quality control module, What-if-game module, data service module, data handler module, and visualization module; the interactions of these components are highlighted in Figure 7.7 (left).

The tool UI has two different modes, the first one presents data following a production stations layout sorted order, displaying the several KPIs for each one of them. The target and the actual KPI values are shown, as well as their evolution trend (positive or negative). A color scheme enriches the visual experience, enabling the user to quickly detect problematic situations. Additionally, a control chart for each KPI is accessible by opening a new UI perspective, displaying its timed evolution.

The what-if-game mode functionality presents data in a spider diagram, aggregating all the relevant KPIs into one graphical display, see

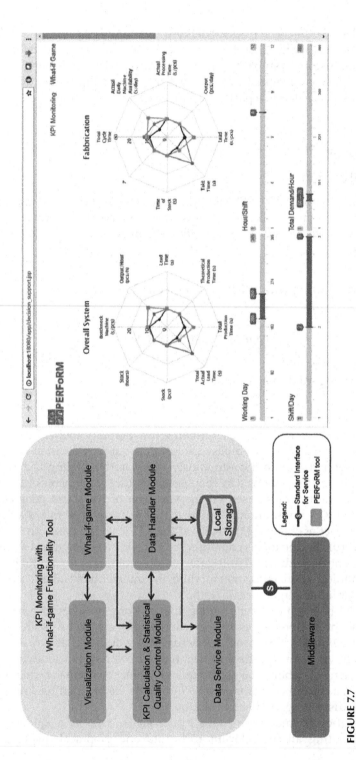

FIGURE 7.7

Architecture of the KPI monitoring with the what-if-game functionality tool (left); view of the tool in the What-if game mode (right). (From Pereira et al. 2017.)

Figure 7.7 (right). The left side of the figure shows the what-if KPI results of the overall system, while the right part of the figure shows other levels of granularity by searching KPIs at the processing/station level.

7.4.2 Universal Web-Based KPI Visualization

TARGET DEFINITION

A Web-based platform for visualization of decision support tools with respect to flexible and reconfigurable production systems.

To provide a generic tool for visualization and monitoring of project outcomes, a universal visualization tool was developed. The visualization tool development aims on an easy-to-understand and intuitive interface that is platform independent and applicable on mobile devices. To accomplish this goal a JavaScript- and HTML-based GUI is proposed. To meet the requirements of the different project outcomes and final solutions during the project duration and supporting the transformation to flexible production systems, the visualization is based on a modular assembly principle. All elements within the UI are defined as entities. Entities can serve various tasks and be adapted to the needs of the user. An entity can, for instance, be a machine, a process, a whole factory, an analytic method, a simulation, or a whole SE. A hierarchic structure can be easily implemented and has a dynamic behavior. Hence, dependencies between production facilities and reconfiguration of an agile production system can be presented. Changes in dependencies between production facilities and entities are dynamically updated. Figure 7.8 (right) shows a prototypical implementation of the universal visualization application. For example, can the root view include all machines to be monitored on the first level? Every machine has tiles for presentation of one outcome (e.g., the monitorable components from the solution presented in Section 7.3.1 is shown in colors representing the current component condition. Another tile shows the date of the last test run. The tiles for the component condition and for the last test run are clickable. The root view is presented in Figure 7.8 (right).

In order to realize communication to the PERFoRM Middleware component a REST to WebSocket Adapter is used. WebSocket is generally chosen as the main communication protocol to have a naturally full-duplex and bidirectional connection allowing secure data transfer (Wang, Salim, and Moskovits 2013). That clears the way to present live data within the visualization. Figure 7.8 (left) depicts the structural approach for the transfer of configuration and visualization data between the Middleware and the visualization. The WebSocket connection is used for a first subscription to the

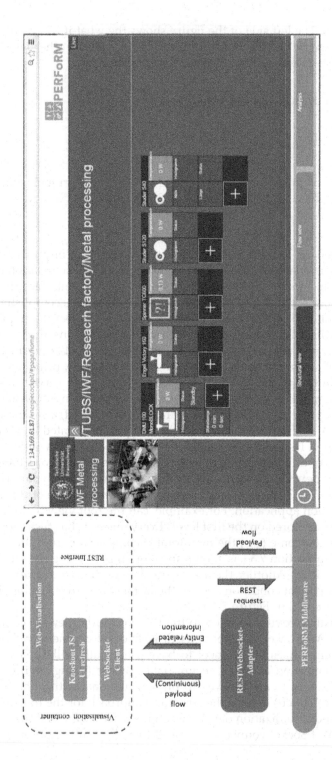

FIGURE 7.8
Communication to the visualization container (left); graphical user interface of the universal Web-based visualization (right). (From EU HORIZON 2020 FoF PERFoRM 2017.)

Middleware to define and set the relations between the entities. The data transfer to the entities is then realized either via the WebSocket connection (e.g., for a continuous data stream) or an http-REST connection.

7.5 Planning Logic

The PERFoRM approach aims to develop solutions toward the integration of industrial agents for scheduling and reconfiguration of manufacturing systems and robot cells. Therefore, hybrid approaches were developed introducing industrial agents in a combination with optimization approaches.

7.5.1 Hybrid Agent-Based/Optimization-Based Scheduling Approach

TARGET DEFINITION

Development of a hybrid scheduling approach for flexible and reconfigurable manufacturing systems. Combining the advantages of centralized optimization algorithm with the advantages of a distributed agent-based approach.

Both centralized and distributed agent-based planning and scheduling systems have unique pros and cons. In agent-based scheduling, a greedy strategy is used to make a locally optimal decision at each stage of the process with the hope that a globally optimal solution could emerge. A greedy strategy does not always produce an optimal solution because of its shortsightedness and non-recoverability. Agents can make certain commitments too early in the execution process, which prevent them from finding the best overall solution later. In order to create a system that is close to optimal and also robust to disturbances, the traditional centralized method is combined with the agent-based approach.

As shown in Figure 7.9 (left), a genetic algorithm (GA) is used to generate an initial optimized schedule. The initial schedule is then used to generate a dispatch rule for the agents during the workflow execution phase. A dispatch rule known as the earliest operation due time (EODT) is created from the schedule. The EODT is different from the conventional earliest due time dispatch rule. It represents the start time of an operation in the optimized schedule rather than the overall due time of the part. In terms of implementing the proposed solution on the shop floor, the manufacturing execution system (MES) layer will need to be structured differently from the existing architecture. The architecture that will support the solution is shown in Figure 7.9 (right).

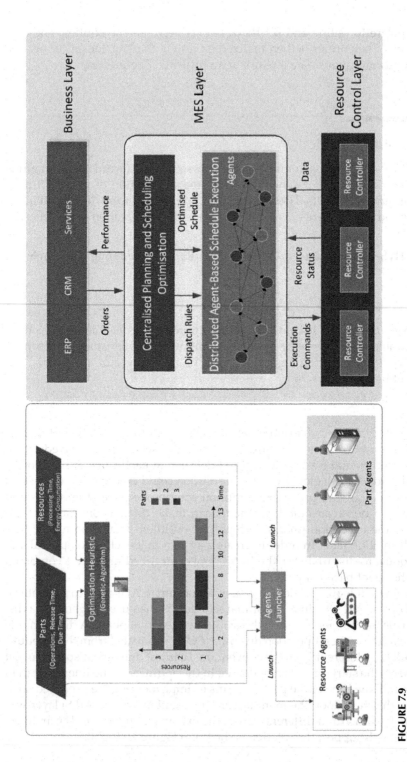

FIGURE 7.9

Hybrid centralized and agent-based scheduling system (left); MES architecture for the hybrid scheduling system (right). (From Ogun et al. 2018.)

7.5.2 Hybrid Agent-Based/Optimization-Based Reconfiguration Approach

TARGET DEFINITION

Development of a hybrid reconfiguration mechanism for flexible and reconfigurable manufacturing systems. Combining the advantages of optimization algorithms with the advantages of an agent-based approach.

The *hybrid agent-based/optimization-based reconfiguration approach* extends the idea of the scheduling approach by adding the functionality of reconfiguration. Reconfiguration planning in concurrent manufacturing of parts from multiple families is an NP-hard combinatorial optimization problem. The GA is effective in solving such problems, especially when the amount of input data is large, and the rate of exponential data growth for discrete optimization grows linearly. Although the GA cannot guarantee a globally optimum solution in polynomial time, it is able to provide near-optimal solutions within a shorter space time compared to other deterministic algorithms for search space optimization. One of the characteristics of reconfigurable manufacturing systems is convertibility, which is the ability to quickly adjust the functionality of the system and controls to suit new production requirements. However, decisions regarding how to deal with exceptions in the reconfiguration process are complex. Multi-agent systems have shown great potentials in dealing with dynamic situations. Therefore, agents are used to manage the reconfiguration schedules produced by the optimizer because complex decisions can be made through the synergy resulting from the reasoning and negotiation mechanisms of the agents (Figure 7.10, left).

Although it has been proven that agent-based schedulers provide better overall performance in environments that require reconfigurability, flexibility, and reliability, the technology has not gained much industrial adoption and deployment. This is due in part to the lack of support for distributed systems in current MESs and manufacturing operation management (MOM). In order to facilitate the adoption of agent-based technology, a new structure for linking the reconfiguration planning and scheduling layers of production systems (MOM or MES) to higher- and lower-level layers needs to be defined. The proposed framework for linking MOM to the enterprise business logic and control layers is shown in Figure 7.10 (right).

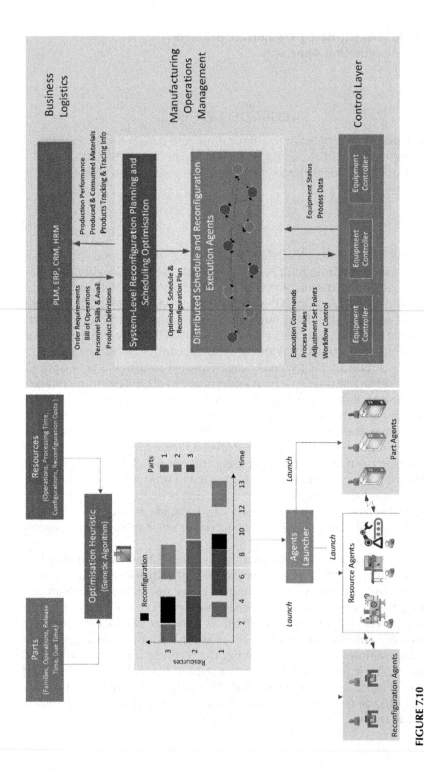

FIGURE 7.10
Hybrid centralized and agent-based reconfiguration planning and execution system (left); framework for linking manufacturing operation management (MOM) to business logic and control layers (right). (From EU HORIZON 2020 FoF PERFoRM 2018.)

7.5.3 Reconfiguration Mechanisms on the Machine Level

TARGET DEFINITION

Development of an agent-based reconfiguration approach on the machine level.

The agent-based reconfiguration focuses on the logical reorganization of the reconfigurable production cells. The reconfiguration is triggered according to the schedule received from the planning and scheduling logic. The process modules are plugged and unplugged to fulfill the work orders defined in the schedule. When an unpredictable event occurs, decisions are made by the agents in real time to establish corrective mechanisms for maintaining the stability and feasibility of the pregenerated schedule.

The process agents (see Figure 7.11, left) are created according to the catalogue of modular process modules that can be used in the production cell, creating one process agent for each physical process module. After the initialization phase, where the agent is parameterized according to the details of the process module it represents, namely its name and the Open Platform Communications Unified Architecture (OPC-UA) server address, and subscribes some process's parameters in the OPC-UA server, the agent enters in a sleep mode, waiting for an event (as result of the initial subscription) notifying that its associated process is plugged-in. At this stage, the agent switches to an active mode, updates its position and location within the micro-flow production cell, and notifies the robot agent responsible for this cell about its new state, providing information regarding the process and its position. The same procedure is executed when the process agent receives an event notifying that the process is plugged-out. This mechanism allows the seamless reconfiguration of the cell in a distributed manner, since the process agents are individually detecting when are plugging-in and -out and exchanging information with each other and with the robot agents to maintain the proper knowledge about the cell configuration. In active mode, each process agent is also continuously collecting data related to the performance of its process module; for example, processing time, number of processed parts, and status (idle or execution). This data is analyzed and aggregated and sent to an external KPI monitoring tool.

The robot agent (see Figure 7.11, right) is launched at the same time as the process agents, creating one robot agent for each physical robot placed in the micro-flow production cell. Usually, a micro-flow cell has only one robot and only one robot agent is created, but if a given system is composed by several micro-flow cells, then more robot agents will be created. After the initialization phase, similar to the one described for the process agent, where the agent is parameterized according to the robot type, the agent remains in active

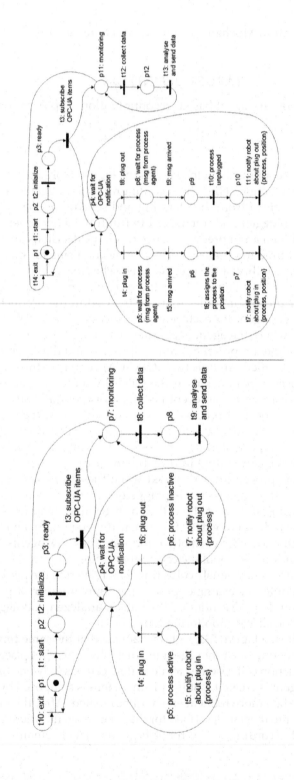

FIGURE 7.11

Behavior model for the process agent (left); behavior model for the robot agent (right). (From Dias et al. 2017.)

mode, monitoring the performance of its physical counterpart and waiting for an event, as result of the initial subscription: A plug-in or plug-out of a modular process. When a process module is plugged-in, the robot agent is automatically notified about the change in the cell configuration, storing the info related to the plugged-in process and its location in its local database. This information is assigned to the robot controller that should adapt itself to work with this new modular process at this location, namely downloading the proper robot program and changing its tools. A same procedure occurs when the robot agent receives an event notifying that a process is plugged-out. In this case, the agent updates the unavailability of the process and informs the robot controller that this process is not anymore active. After the reconfiguration of the micro-flow production cell, the cell can return to the processing state, if the cell safety procedures are guaranteed. In fact, aiming to ensure the safety, robot agents should check the information related to the cell safety systems, for example, light curtains and emergency stops, before allowing to return to the normal processing operation.

7.6 Conclusion

This chapter described a selection of different methods, tools, and services developed following the PERFoRM approach. The integration and use of the individual solutions is described in the later chapters addressing different industrial Use Cases.

References

Anylogic. 2018. "The AnyLogic Company." Online available via https://www.anylogic.com/, last verified at October 10, 2018.
AutomationML. 2014. "Whitepaper AutomationML Part 1—Architecture and general requirements." Online available via https://www.automationml.org/o.red.c/dateien.html, last verified at October 10, 2018.
Barga, R., V. Fontama, W. H. Tok, and L. Cabrera-Cordon. 2014. "*Predictive Analytics with Microsoft Azure Machine Learning.*" Chapter 4, pp. 67–83. Apress.
Büth, L., N. Broderius, C. Herrmann, and S. Thiede. 2017. "Introducing Agent-Based Simulation of Manufacturing Systems to Industrial Discrete-Event Simulation Tools." In *proceedings of Industrial Informatics (INDIN), 2017 IEEE 15th International Conference on*, pp. 1141–1146. Emden, Germany.
Chakravorti, N., M. M. Rahman, M. R. Sidoumou, N. Weinert, F. Gosewehr, and J. Wermann. 2018. "Validation of PERFoRM reference architecture demonstrating an application of data mining for predicting machine failure." *Procedia CIRP.* 72: 1339–1344.

Dias, J., J. Vallhagen, J. Barbosa, and P. Leitão. 2017. "Agent-Based Reconfiguration in a Micro-Flow Production Cell." In *Industrial Informatics (INDIN), 2017 IEEE 15th International Conference on*, pp. 1123–1128. IEEE.

DIN/VDE. 2017. "DIN EN 62890:2017-04." Accessed 10, 2018. https://www.beuth.de/de/norm-entwurf/din-en-62890/269121145.

EU HORIZON 2020 FoF PERFoRM. 2017. "Deliverable D4.3: Decision rules and KPI and Functionality Visualizations." Unpublished manuscript, last modified 10, 2018. http://www.horizon2020-perform.eu/index.php?action=documents.

EU HORIZON 2020 FoF PERFoRM. 2018. "Deliverable D4.4: Guidelines and Recommendations for Reconfigurability Mechanisms for Machinery and Robots." Unpublished manuscript, last modified October 10, 2018. http://www.horizon2020-perform.eu/index.php?action=documents.

"EU HORIZON2020 FoF PERFoRM 2015-2018." Accessed October 10, 2018. https://cordis.europa.eu/project/rcn/198360_de.html.

Fischer, Jan, Birgit Obst, and Benjamin Lee. 2017. "Integrating Material Flow Simulation Tools in a Service-Oriented Industrial Context." In *2017 IEEE 15th International Conference on Industrial Informatics (INDIN): University of Applied Science Emden/Leer, Emden, Germany, 24-26 July 2017: Proceedings*, pp. 1135–1140. Piscataway, NJ: IEEE.

IEC. 2017. *Smart manufacturing - Reference architecture model industry 4.0 (RAMI4.0): 1.0th ed. 25.040.01 - Industrial automation systems in general 35.080 - Software 35.240.50 - IT applications in industry*, no. 63088:2017: IEC. https://webstore.iec.ch/publication/30082.

McKinsey & Company. 2015. "Industry 4.0: How to navigate digitization of the manufacturing sector." Accessed October 10, 2018. http://worldmobilityleadershipforum.com/wp-content/uploads/2016/06/Industry-4.0-McKinsey-report.pdf.

Neef, B., J. Bartels, and S. Thiede. 2018. "Tool Wear and Surface Quality Monitoring Using High Frequency CNC Machine Tool Current Signature." In Proceedings *Industrial Informatics (INDIN), 2018 IEEE 16th International Conference on*, pp. 1045–1050. Porto, Portugal.

Ogun, P., N. Lohse, L. Büth, S. Thiede, and S. Forsman. 2018. "Planning and scheduling model for a flexible semi-dynamic combination layout production plant." *In Press: IEEE Transactions on Industrial Informatics*.

Pereira, A., P. Petrali, A. Pagani, J. Barbosa, and P. Leitão, 2017. "Dynamic Monitoring of Key-Performance Indicators in Industrial Environments." In *Industrial Informatics (INDIN), 2017 IEEE 15th International Conference on*, pp. 1129–1134. Emden, Germany.

Shin, H-C. et al., 2016. "Deep convolutional neural networks for computer-aided detection: CNN architectures, dataset characteristics and transfer learning." In *Transactions on Medical Imaging*. 35 (5): 1285–1295. IEEE.

Siemens. 2018. "Tecnomatix Plant Simulation." Online available via https://www.plm.automation.siemens.com/store/en-us/plant-simulation/last verified at 10, 2018.

Wang, V., F. Salim, and P. Moskovits. 2013. *The Definitive Guide to HTML5 WebSocket*. Vol. 1. New York: Apress.

Yanai, K., and Y. Kawano. 2015. "Food Image Recognition Using Deep Convolutional Network with Pre-Training and Fine-Tuning." In Proceedings *2015 IEEE International Conference on Multimedia & Expo Workshops (ICMEW)*.

8

Migration Strategy toward Innovative, Digitalized, and Harmonized Production Systems

Ambra Calà

(Siemens AG)

Ana Cachada, Flávia Pires, José Barbosa

(Instituto Politécnico de Bragança)

Jeffrey Wermann, Armando W. Colombo

(Institute for Informatics, Automation and Robotics (I²AR))

CONTENTS

8.1 Introduction

Industry today is facing the fourth Industrial Revolution, also called Industry 4.0, pulled by the market demand of shorter delivery time and product life cycles, with increased product variety and smaller lot sizes, and pushed by innovative technologies, such as Cloud computing, big-data analysis, connected Cyber-Physical Systems (CPSs), Internet of Things, and related services, as well as autonomous robots and augmented reality. Industry 4.0 is an industrial paradigm that is proposing to transform the traditional factories into smart factories, which are more competitive, efficient, and productive. The goal is the rapid introduction of new tangible products and intangible products (services) into the market as soon as market and customer requirements change. In order to achieve this goal, the production process itself should be more agile and particularly correctly digitalized.

Several smart technologies, intelligent systems, and innovative solutions have been developed in the past years to enable the realization of digitalized production systems. However, the introduction of new technologies in existing production environments is always very difficult. Manufacturers, in order to accomplish their goal of digitalization, need to be supported with established methodologies to migrate toward innovative systems, namely those applying Cyber-Physical Production Systems (CPPSs) concepts, starting from the key element of this change: the automation control architecture. As described in Chapter 3, Production harmonizEd Reconfiguration of Flexible Robots and Machinery (PERFoRM) developed a reference architecture that aims at decentralizing the conventional automation architectures, based on the ISA-95 automation pyramid (ISA Organization 2018), in order to enable the interaction and interoperability of heterogeneous devices toward a more flexible and reconfigurable production system (Givehchi et al. 2017).

However, the implementation of new control architecture and smart technologies has a big impact in existing industrial environments, considering the current legacy systems and processes. Therefore, an important topic within the framework of Industry 4.0 is the migration toward innovative production systems. A smooth migration strategy from existing traditional centralized systems toward digitalized, harmonized, flexible, and reconfigurable production systems is necessary to mitigate risks and support the industrial adoption of Industry 4.0 technologies.

According to ARC Advisory Group (2007), the decision to perform a system migration has different triggering sources:

- New business opportunities become impossible to accomplish without a new system.
- The system is no longer cost-effective to support.

- The system is inflexible and doesn't respond to customer demands.
- The system lacks visibility that could prevent equipment breakdown and disruption in the supply chain.
- The system is impossible to be expanded.

In the advent of Industry 4.0, the need to implement the new innovative, digitalized, and harmonized production systems also constitutes an opportunity to trigger the migration process.

8.2 State of the Art

Over the last decade various migration approaches, in terms of strategies and processes, have been developed in different research domains to enable the transformation of an existing system into a new one. This chapter analyzes the migration strategies and processes that can be found in literature.

8.2.1 Migration Concepts and Approaches

A process is defined by Transvive (2011) as a logical sequence of tasks performed to achieve a particular objective. In the context of migration, the process should clearly define a set of activities to be carried out in order to successfully transform a considered system. The migration processes found in literature mainly focus on the transformation of legacy information systems or on data migration, while most recent processes addressed the migration of legacy applications into services and cloud computing.

The first model describing a migration process is the Butterfly Methodology, defined by Wu et al. 1997. This generic process consists of five phases: (i) justification, (ii) legacy system understanding, (iii) target system development, (iv) migration, and (v) testing. Following this stepwise approach, the legacy system is fully replaced by the target system only after validation during the testing phase, enabling the rolling back of the legacy systems at any stage and reducing the risks related to the implementation of new systems. This generic process founds the basis for more specific migration processes, such as the following processes addressing the migration toward Web services: SMART, SOAMIG, MASHUP, and IMC-AESOP (Colombo et al. 2014) approaches.

The Service Migration and Reuse Technique (SMART) provides an iterative process for migrating legacy IT systems to services in a Service-Oriented

Architecture (SOA), which consists in six activities: (i) establish context, (ii) define candidate services, (iii) describe existing capabilities, (iv) describe target SOA environment, (v) analyze the gap, and (vi) develop strategy (Lewis et al. 2006). The goal of the process is to help decision-makers in analyzing the feasibility, risks, and involved costs of the target solution.

The SOAMIG (MIGration of legacy software into Service-Oriented Architectures) project also mentions the development of a migration process toward SOA, which is developed as an iterative process and is represented by four phases: (i) preparation, (ii) conceptualization, (iii) migration, and (iv) transition (Ionita, Lewis, and Litoiu 2013). This migration process model emphasizes the transformation-based conversion toward SOAs addressing also data, code, and user interfaces.

The MASHUP (MigrAtion to Service Harmonization compUting Platform) (Cetin et al. 2007) is another technique addressing the migration of legacy systems into service-oriented computing. This migration process proposes six steps: (i) model, (ii) analyze, (iii) map and identify, (iv) design, (v) define, and (vi) implement and deploy. This roadmap addresses both the behavioral and architectural aspects of the migration process toward the targeted mashup technology, on which the service-oriented computing is based.

The work developed in the ArchitecturE for Service-Oriented Process—Monitoring and Control (IMC-AESOP) project (EU FP7 AESOP Consortium 2010–2014) is mainly focused in the implementation of SOA to change the existing systems into distributed and interoperable systems (Colombo et al. 2014; Delsing et al. 2012). The migration of systems toward SOA has four major steps: (i) initiation, (ii) configuration, (iii) data processing, and (iv) control execution. Also, the migration process makes use of mediator technology to communicate with the legacy systems, that is, the old systems. The four steps have been designed to maintain the perception of conformity between the several interfaces.

Differently to the previous approaches the Cloudstep focuses on the migration of legacy applications to the Cloud (Beserra et al. 2012). It is a step-by-step decision process based on nine activities to identify and analyze the impact aspects that can influence the selection of the cloud solution and also the migration tasks. These activities are (i) define organization profile, (ii) evaluate organizational constraints, (iii) define application profile, (iv) define cloud provider profile, (v) evaluate technical and/or financial constraints, (vi) address application constraints, (vii) change cloud provider, (viii) define migration strategy, and (ix) perform migration.

Finally, the XIRUP (eXtreme end-User dRiven Process) process addresses the iterative modernization of components-based systems (Fuentes-Fernández, Pavón, and Garijo 2012). The process has been developed in the MOMOCS (MOdel driven MOdernisation of Complex Systems) project and comprehends four stages: (i) preliminary evaluation, (ii) understanding, (iii) building, and (iv) migration. The goal of XIRUP is not only to provide a sequence of activities to migrate a legacy system into a new one but also to guide the users in evaluating possible migration alternatives based also on cost-benefit analysis.

The migration processes described above present some similarities despite their different goals. In general, they follow a stepwise approach that starts with the identification of the legacy system and the definition of the target solution of the migration based on the requirements. It continues with the development of the target system and the definition and performance of the migration. The definition of the target solution is based only on technical constraints, like in SOAMIG and IMC-AESOP, or includes also business requirements, like in SMART, MASHUP, and XIRUP, or even legal, administrative, and organization constraints, like in Cloudstep.

However, these processes address a very specific technical solution, for example SOA, and cannot be adopted, as they are, in the evolving context of Industry 4.0. For the implementation of such a new business paradigm, it is necessary to have a migration process that allows for continuous improvement.

8.2.2 Migration Strategies

Nowadays, there are three main migration strategies present in the literature, namely the Big Bang, Parallel Systems, and Phased. Although general and normally applied to the migration of software, they can also be adapted to the CPPS migration. Each is composed by several stages, which includes the assessment of the current environment to migrate (legacy system or As-Is), planning for the development of a migration project, architecting a new target environment (target system or To-Be), implementing a migration by using available tools and processes, and managing the newly migrated environment (Transvive 2011). Also, crucial steps are the analysis of the risks involved and the development of a contingency plan in case of migration failure (Cala et al. 2017).

Briefly, the three strategies can be summarized as (a deeper explanation can be found in Cachada et al. (2017)):

- The Big Bang strategy can be described as a change in a single moment in time, switching off the legacy system and switching on the target system, on a designated date (Critical Manufacturing 2015), known as the Go-Live date (Madkan 2014).
- The Parallel Systems strategy is characterized by both legacy and target systems run at the same time, that is, in parallel, for a certain period of time (Critical Manufacturing 2015; Open Text 2009). This time corresponds to the migration execution time, in which the legacy system is designated as Master and the target system as Slave. The target system becomes the Master system only after it is tested and validated, and then, the legacy system becomes the Slave system or is switched off. If the legacy system continues running as Slave, additional costs need to be considered in the migration process (ARC Advisory Group 2007).

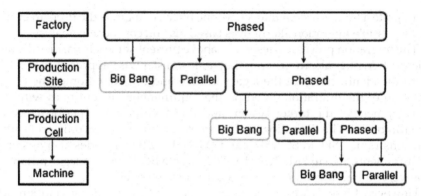

FIGURE 8.1
Recursivity in the implementation of the Phased Strategy. (From Cachada et al. 2017.)

- The Phased Strategy allows executing the migration through a gradual transition, following a well-planned sequence (Lewis et al. 2006), which requires an intensive study of interdependencies and processes' priorities in order to know the correct sequence of the migration phases.

An important aspect of this migration strategy is the definition application areas, followed by a definition of the secondary types of the migration strategies. In fact, for each phase, the previously described strategies, that is, Big Bang, Parallel, and also the Phased strategy, can be used independently. This represents a recursivity in the implementation of this strategy, meaning that it is possible to repeat recursively the choice of the migration strategies, namely Big Bang, Parallel, and Phased strategies, according to the granularity of the factory level, as illustrated in Figure 8.1.

In this case, and selecting the Phased strategy for the migration at the factory level, the gradual migration at the production sites can be implemented by considering Big Bang, Parallel, and/or Phased strategies for the different sites. If a Phased strategy is selected for one production site, its implementation at production cell level can recursively use the same approach, being implemented by using Big Bang, Parallel, and/or Phased strategies for the different cells (Cala et al. 2017).

This process is mainly used by large corporations that will build and test a core solution with common functionality and processes, before applying it as part of the Phased solution. For smaller corporations, it may not be useful to compromise to a highly staggered plan (HSO 2016).

The selection of the best migration strategy to be adopted depends on the environment and the addressed technical, economic, and social conditions. The comparison of the different strategies considers the assessment of several features, such as risks, migration design time, migration execution time, downtime, and costs (effort), as summarized in Table 8.1.

Briefly, it is possible to conclude that the Big Bang strategy has a low implementation cost but involves a higher risk, migration design time, and downtime.

TABLE 8.1

Comparison of the Migration Strategies

	One-Shot	Parallel	Phased
Risk	HIGH	LOW	MEDIUM
Migration design time	HIGH	LOW	MEDIUM
Migration execution time	LOW	MEDIUM	HIGH
Downtime	HIGH	LOW	MEDIUM
Cost (effort)	LOW	HIGH	MEDIUM

Source: Critical Manufacturing (2015).

In opposite, the Parallel strategy has a low risk, migration design time, and downtime but represents a high cost for the company. The Phased strategy is a kind of compromise between these two approaches, presenting the highest migration execution time.

8.3 The PERFoRM Migration Approach

The PERFoRM Migration approach (Figure 8.2) is a stepwise approach that can support manufacturers in minimizing the risk related to the implementation of Industry 4.0 solutions, especially in this evolving context in which the way forwards, the target conditions and the available technology options are still unclear and uncertain. The approach consists in breaking down the migration path toward the long-term vision of Industry 4.0 in intermediate goals aiming at implementing the target system step by step. In fact, by analyzing and performing the migration steps one at time, it is possible to smoothly transform and improve a production system toward Industry 4.0.

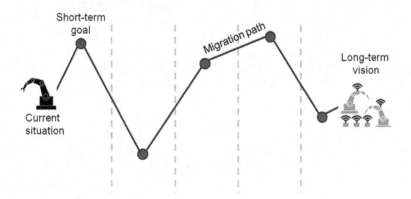

FIGURE 8.2
PERFoRM stepwise migration approach.

8.3.1 Migration Process

In order to support the continuous and incremental improvement of a production system toward innovative production systems, a five-phase migration process (Figure 8.3) has been developed within PERFoRM to support the identification, design, and implementation of each migration step by means of adequate methods and tools, described in the following section (Cala et al. 2017).

The process starts with the Preparation phase in which the existing system is analyzed and the possible target system is defined as a long-term vision of the migration and the impact aspects to be considered. In this phase, first the analysis of the current system, that is, the available equipment, technology systems, and applications in the production system, as well as control automation infrastructure and communication network, is performed (*context analysis*). Second, the goal of the migration is defined based on the manufacturer's motivation that is leading to the migration (*goal definition*).

The following Options Investigation phase aims at investigating possible migration solutions based on the As-Is and To-Be situations analyzed in the previous phase. First, existing and available technology systems and applications are collected (*collection*), investigated, and then the optimal one is selected according to the manufacturer's needs (*selection*). The selection of the optimal migration solution step to achieve the short-term goals and thus the target system in the end depend on the relevant impact aspects identified in the previous phase.

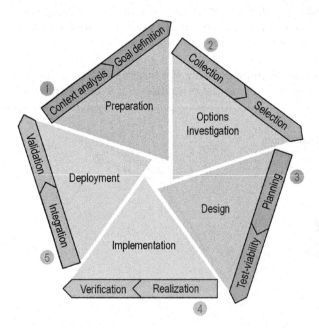

FIGURE 8.3
PERFoRM smooth migration process. (From Cala et al. 2017.)

The selected solution option is then detailed in the Design phase. Here the concept and design of the components of the target system, that is, new tools, adapters, and interfaces, are defined, as well as their integration with the legacy systems (*planning*). The viability of the solution is also tested in this phase (*test-viability*) to ensure that the next phase is not compromised by a faulty design solution. If the results of the viability tests do not match with the expected benefits, the user can repeat the previous phases and select a different option or re-define the goal of the migration.

In the Implementation phase, the designed solution is realized and verified. Since the design of the migration plan was proven in theory, it is going to be put in practice here (*realization*). The plan will be followed according to the selected migration strategy, implementing all the required technologies, supporting the transition from the legacy system to the new innovative, digitalized, and harmonized production system. Then, every step of the plan is verified in terms of accomplishment and achievement of the expected results (*verification*).

Finally, the Deployment phase is related to the installation of the new system in a real environment state (*integration*) and its further validation to ensure that all the system's qualities are functional and the new system performs as intended (*validation*). After this phase the first migration step is completed and the process can start over again. The process is repeated for each migration step until the achievement of the long-term vision.

The migration path of the manufacturing systems is not a straightforward process. It involves a very complex list of tasks and various steps, including the selection of a migration strategy. The created migration process was built around the migrations strategies, which makes the progression of the migration more secure since it is a stepwise process.

8.3.2 Migration Methods and Tools Based on a Petri Nets Approach

The PERFoRM migration process and its phases are described in Figure 8.4 using Petri nets (Cachada et al. 2017; Colombo, Carelli, and Kuchen 1997; David and Alla 1994; Leitao, Colombo, and Restivo 2006; Zhou and DiCesare 1993). Preparation and Options Investigation are represented in transitions $t2$ and $t3$, while Design, Implementation, and Deployment are described according to the three implementation approaches identified in the state-of-the-art, namely Big Bang, Parallel Systems, and Phased, which correspond to the One-Shot ($t9$), Parallel ($t8$), and Phased ($t7$) strategies.

Once the migration plan is successfully implemented and the commissioning of the target system has been achieved, the migration process is completed; therefore, the target system is ready to run (place $p9$). This process can be cyclical, meaning that a new migration process can be initiated if necessary and the environment that was defined as "target system" is now the "old system."

The following sections detail the modeling of each of the defined phases.

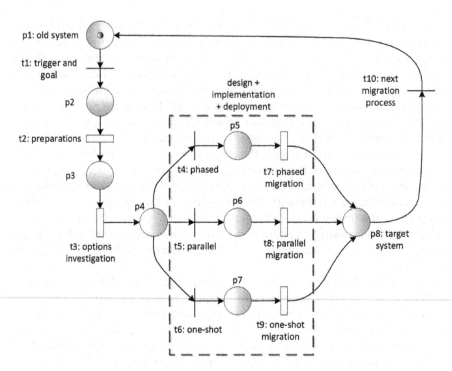

FIGURE 8.4
Petri nets model for the PERFoRM smooth migration.

8.3.2.1 Modeling the Preparation Phase

The Preparation phase is the first phase to be executed in the migration process. The Petri nets model for the Preparation phase, illustrated in Figure 8.5, starts with the identification of the business vision, which defines the direction of the migration in the long run. The following tasks are related to the assessment of the legacy system and the identification of the possible migration opportunities in the short-medium run (transitions $t2.t3$ and $t2.t4$, respectively).

The assessment of the current system intends to lead to a complete comprehension of the system and to identify what needs to be changed. In parallel, the assessment of the migration opportunities is also performed, where several details are defined, such as the motivation to execute the migration and what actors and systems are involved in the migration process. The assessment of the migration opportunities also includes the evaluation of several scenarios for the target system. Based on these assessments and coherently with the identified business long-term vision, the next target condition of the system is defined.

The target condition of the system represents the next concrete short-term goal to be achieved in the direction of the long-term vision. Then, a preliminary risk and impact analysis is performed (represented by transition $t2.t6$).

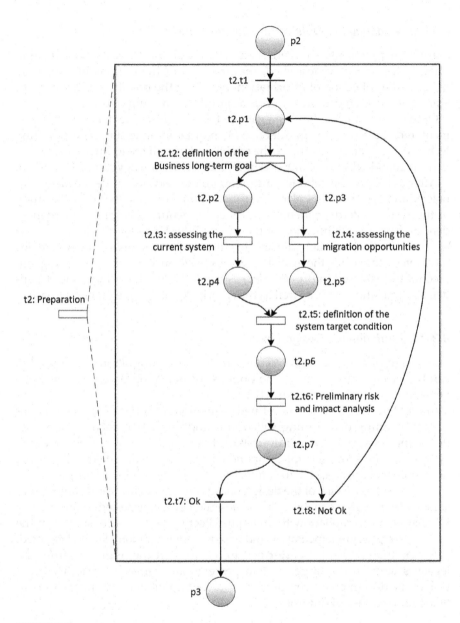

FIGURE 8.5
Petri nets model for the Preparation phase.

This analysis intends to verify if the desired migration presents risks that can be mitigated or instead if the migration is too dangerous to be performed. In the second case, it is necessary to start a new iteration in the preparation process. On the other hand, if the identified risks are admissible, the Preparation phase is completed.

8.3.2.2 Modeling the Options Investigation Phase

Figure 8.6 represents the Petri nets model for the Options Investigation phase. As previously described, the first task of the options investigation phase is the collection of information regarding the possible solutions (transition $t3.t2$), which usually requires a significant amount of time.

Once the information has been gathered, it is necessary to make their assessment (represented by the transition $t3.t3$) in order to understand if the existing technology is suitable to achieve the established goal. If the technology does not fit the demands, the possibility of developing new technology should be studied.

After assessing the collected information, the selection of the needed technology and systems takes place (transition $t3.t6$). This is followed by the study of the system interdependencies in order to identify the critical interdependencies. The misidentification of the critical interdependencies, in a stepwise approach for the migration, can lead to failure in the migration process. Finally, the strategy to execute the migration is selected (transition $t3.t11$) by taking into account the assessed information (transition $t3.t3$) and the critical interdependencies (transition $t3.t7$), as well as the criteria defined in the previous phase.

8.3.2.3 Modeling the Design Phase

The design phase is common to the three different migration strategies that can be selected at the end of the options investigation phase, and its Petri nets model is depicted in Figure 8.7.

Initially the target system is defined (transition $t4.t2$), namely its functionalities, for example data mining, scheduling and simulation, set of components (tools and legacy systems), information flows, and connection with legacy systems. Second, the several components previously identified, such as the new tools and the adapters to connect the legacy systems, are designed.

The second main task of the design phase is related to the test viability (transition $t4.t7$), where the viability of the designed system is tested to understand if the system is compliant with all required activities. In case of success in the viability test, the implementation and deployment migration plan is elaborated.

A risk analysis is also performed to ensure that the migration from the legacy system to the target system presents an admissible risk. The final task of the design phase comprehends the definition of a contingencies plan (to be used in case of failure).

8.3.2.4 Modeling the Implementation and Deployment Phases

The Implementation and Deployment phases are strongly dependent on the migration strategy selected during the Options Investigated phase for the migration process. Next sections will detail these phases under the perspectives of the three different migration strategies considered in the migration approach designed as part of the PERFoRM approach.

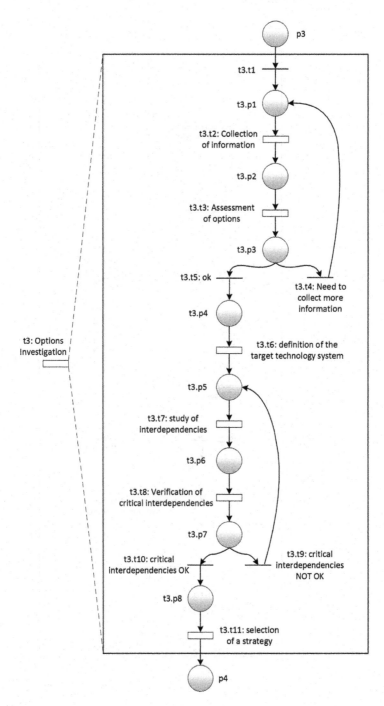

FIGURE 8.6
Petri nets model for the Options Investigation phase.

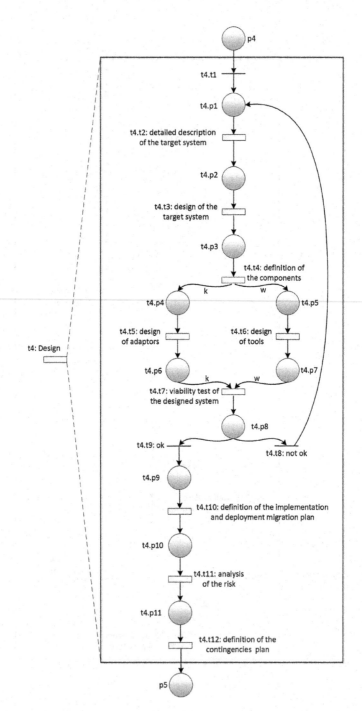

FIGURE 8.7
Petri nets model for the "design phase" transition.

8.3.2.4.1 One-Shot Strategy

The One-Shot strategy was inspired in the Big Bang strategy, where all the changes to be executed happen in a single period of time, that comprises the time to uninstall the old system and the time to install and validate the target system. The application of this strategy requires that the target system has to be completely defined and validated off-line. With this system ready, the old system is switched off and the target system is deployed as an integrated solution, being commissioned only if successfully validated.

This strategy, broadly used, for example in automotive industry, represents a high risk for the company since the old system is shut down, which makes almost impossible to rollback.

Analyzing the Petri nets model for the migration process (see Figure 8.4), the One-Shot migration strategy is performed when the transition *t10* is fired, which can be exploded into a sub-Petri nets model represented in Figure 8.8.

This migration strategy comprises the execution of a sequence of steps that starts with the development of the necessary system components based on new technologies or paradigms. After this stage, the system is ready to be deployed in the factory and the original system can be switched off (transition *t10.t2*).

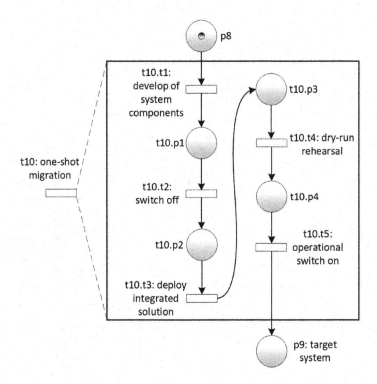

FIGURE 8.8
Petri nets model of the One-Shot migration strategy.

Once the old system is shut down, the integrated solution is deployed and a dry-run rehearsal is performed to certify that the target system is ready to run (transition *t10.t4*). When the dry-run rehearsal is successfully completed, the system is switched on (*t10.t5*) and the migration project is commissioned.

Some timed transitions of the Petri nets model can be also exploded to introduce more control details. As example, Figure 8.9 illustrates the sub-Petri nets model for the transition *t10.t1* that represents the development of system components, introducing particularities related to the PERFoRM environment.

Initially, several actions are performed in parallel, namely the development of k adapters (transition *t10.t1.t2*), installation of f Middleware components (transition *t10.t1.t3*), and development of w new monitoring and analytics tools (transition *t10.t1.t5*) and instantiation of the data model, which are key components in the PERFoRM system. Note that the k, f,

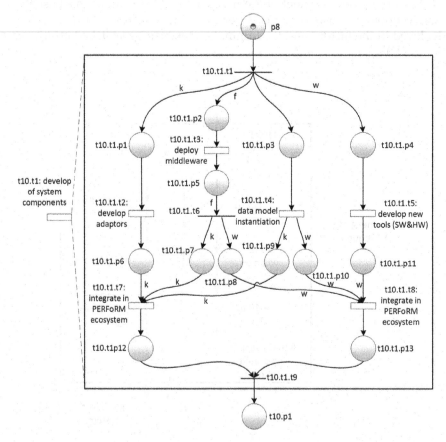

FIGURE 8.9
Petri nets model for the "develop system components" transition.

and w values are defined during the Design phase and are mapped into tokens that populate the places that represent the four referred parallel activities.

Once the entire set of adapters are developed, the data model is instantiated, and middlewares are installed, the legacy systems can be integrated in the PERFoRM ecosystem (transition *t10.t1.t7*). On the other hand, the new tools are integrated in the PERFoRM ecosystem (transition *t10.t1.t8*) once all new tools are developed, the Middleware is installed, and the data model is instantiated. When all these software and hardware components are integrated within the PERFoRM ecosystem, the next tasks of the One-Shot strategy can be performed, as previously described.

The "dry-run rehearsal" activity, detailed in Figure 8.10, is related to the final verification of the migration process.

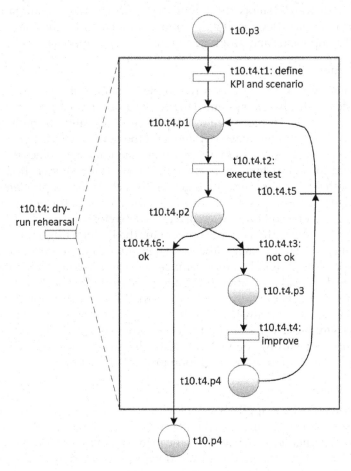

FIGURE 8.10
Petri nets model for the "dry-run rehearsal" transition.

The first step of the dry-run rehearsal is the definition of the test scenario and KPIs (Key Performance Indicators), followed by the execution of a testing (transition *t10.t4.t2*). At the end of the test, two alternatives can happen:

- If "Not Ok," the system needs to be improved and posteriorly be tested again.
- If "Ok," the dry-run rehearsal is approved and the migration project is commissioned and the production resumed.

Note that this sub-Petri nets model can be used by all activities that require the execution of tests.

As previously referred, the implementation of this strategy implies the shutdown of the production site for a period of time. This downtime is strongly dependent on the scope and magnitude of the migration: If the migration only comprises software systems, the downtime is smaller, but if the migration also considers hardware devices, the downtime is higher since the complexity to uninstall components and program and install new components is higher.

8.3.2.4.2 Parallel Strategy

The Parallel strategy is based on the implementation of the target system, side by side, with the old system. This configuration must be kept running in parallel until the target system has proven its viability. Initially, the old system is considered the master system and the new system is the Slave system, but once the target system has proven its viability, it becomes the master system and the old system can become the Slave system or switched off.

Since both systems are running together, the occurrence of problems in the target system (running as slave) is mitigated by the use of the old system and provides a safer period of time to correct its behavior.

Figure 8.11 depicts the Petri nets model for the Parallel migration strategy.

As in the One-Shot strategy, the first steps are related to develop the system components. After all components have been developed, the integrated solution is deployed (transition *t9.t2*), and posteriorly its functionality tested (improving the system if any problem arises). When the new solution is successfully tested and is fully improved, the next step is related to switch on the target system as Slave system and maintain the old system as master. After successfully concluding the viability tests, the target system is switched as master system, finalizing the migration process (transition *t9.t6*).

8.3.2.4.3 Phased Strategy

The Phased strategy is applied by deploying the new system through sequential phases, which requires a well-planned implementation that carefully considers the interdependencies and the priorities of the involved processes.

An important characteristic of this strategy is its recursive nature, meaning that one of the migration strategies can be selected for each phase. As an example, if a Phased strategy is applied to migrate the entire factory, the

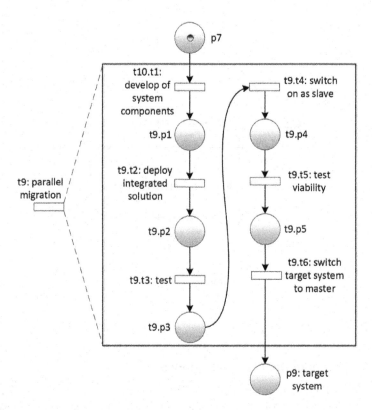

FIGURE 8.11
Petri nets model for the Parallel migration strategy.

migration of each production line can adopt the one-shot, parallel, or Phased strategy, and if this last one is selected, then again one of the migration strategies can be selected for each workstation.

Figure 8.12 illustrates the Petri nets model for the Phased migration strategy. Once the strategy is selected during the design stage, one important note that needs to be taken in consideration is the number of phases and the associated strategy for each one. This information is associated to different variables used to regulate the flow of tokens along the Petri nets model: b represents the number of phases using the One-Shot strategy and p represents those using the Parallel strategy. The number of phased phases is calculated by $L - (p + b)$, where L is the total number of phases.

After selecting this migration strategy, each one of the migration phases is properly executed, considering the defined strategy for each one. A migration phase using the Phased strategy will trigger the recursive application of the same Petri nets model, and migration phases using the One-Shot and Parallel strategies will invoke, respectively, the Petri nets models illustrated in Figures 8.8 and 8.11.

The migration process is concluded when the defined L phases are all successfully implemented and the target system is commissioned.

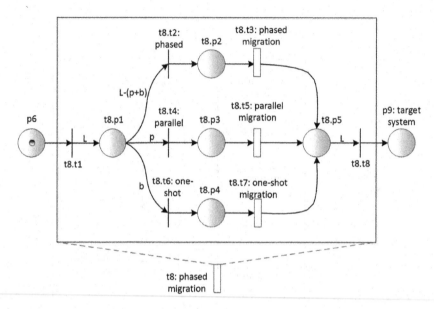

FIGURE 8.12
Petri nets model for the Phased migration strategy.

8.3.3 Validation of the Petri Nets Models

The designed Petri nets models for the implementation of the different migration strategies, for the migration of traditional production systems into digitalized and harmonized production systems, were edited, analyzed, and validated by using the Petri nets Development toolKit (PnDK) (Mendes et al. 2009). In this section, the validation is illustrated by performing a qualitative and quantitative analysis to the general migration process (see Figure 8.4).

The qualitative analysis is related to the structural and behavioral validation of the designed Petri nets models, and particularly the verification of the structural and behavioral characteristics of the model, obtaining information related to the existence of deadlocks, bounded capacity of resources, and conflicts within the system (Feldmann, Schnur, and Colombo 1996). The analysis of the behavioral properties for the Petri nets model representing the general migration process is illustrated in Figure 8.13.

This analysis allows extracting the following conclusions:

- *Safe and 1-Bounded:* The maximum number of tokens that can be in a place is one, which means that only one migration strategy can be selected for the overall migration process.
- *Reversible:* The initial marking is reachable from all reachable markings, which means that after concluding a migration process, a new one can be started if necessary.

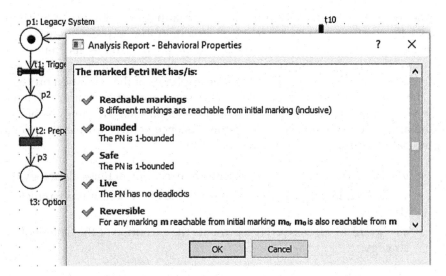

FIGURE 8.13
Behavioral analysis of the Petri nets model.

- *Absence of deadlocks:* For each reachable marking, there is at least one transition that can be triggered to reach another marking, which means that the migration process doesn't stop in any particular step.

Additional characteristics can be extracted through the analysis of the P- and T-invariants, as illustrated in Figure 8.14.

The analysis of the P-invariants allows the verification of mutual exclusion relationships among places, functions, and resources involved in the

✓ **Minimal P-invariants:**

Place	p1	p2	p3	p4	p5	p6	p7	p8	p9
x1	1	1	1	1	1	1	1	1	1

✓ **Minimal T-invariants:**

Transition	t1	t2	t3	t4	t5	t6	t7	t8	t9	t10	t11
y1	1	1	1	1	1	0	0	1	0	0	1
y2	1	1	1	1	0	1	0	0	1	0	1
y3	1	1	1	1	0	0	1	0	0	1	1

FIGURE 8.14
P- and T-invariants of the Petri nets model.

structure and behavior of the model (Feldmann and Colombo 1998; Zhou 1995). For the Petri nets model of the general migration process, there is only one P-invariant, $x1 = \{p1, p2, p3, p4, p5, p6, p7, p8, p9\}$, and by its analysis, it is possible to confirm that only one place can be marked at any time, meaning the mutual exclusion among the several phases of the migration process.

The T-invariants represent the several sequences of operation, that is, the work cycles, exhibited by the behavior model. From the analysis of the T-invariants, it is possible to confirm the existence of three invariants and its physical meaning can be translated as follows:

- $y1 = \{t1, t2, t3, t4, t5, t8, t11\}$ represents the execution of Phased strategy.
- $y2 = \{t1, t2, t3, t4, t6, t9, t11\}$ represents the execution of the Parallel strategy.
- $y3 = \{t1, t2, t3, t4, t7, t10, t11\}$ represents the execution of the One-Shot strategy.

Since the model representing the general migration process comprises several timed transitions that are refined and exploded (see Figure 8.4), the complete analysis of this large model requires the analysis of all sub-Petri nets and the application of the theorems established by Valette (1979) and generalized by Suzuki and Murata (1983) about the preservation of boundedness and liveness properties in Petri nets obtained using the stepwise refinement. The Valette theorem (Valette 1979) states that all properties of a large Petri net can be deduced from the behavioral analysis of the initial Petri net and each one of the sub-Petri nets performed independently.

For this purpose, all timed transitions from the large Petri net, and also the timed transitions included in the exploded sub-Petri nets, were analyzed using the procedures described in, for example Colombo, Carelli, and Kuchen (1997). As an example, the validation of the sub-Petri nets model "develop system components" was performed, as illustrated in Figure 8.15, considering $k = 6, f = 2$, and $w = 4$.

This analysis allows concluding that this model is reversible, absent of deadlocks, and six-bounded (a maximum of six tokens may be hosted in one place, representing the actions to develop six adapters for the identified legacy systems).

Since all sub-Petri nets were validated, concluding that they are bounded and absent of deadlocks, it is possible to conclude that, according to the Valette theorem (Valette 1979), the large Petri nets model for the general migration process is also bounded and absent of deadlocks.

8.3.4 Using Petri Net Tools to Control the Migration Process

The Petri net-based migration process provides the basic methodology and steps to migrate to a production system. To actually make use of such a process, it is necessary to instantiate the process for a specific use case. To do

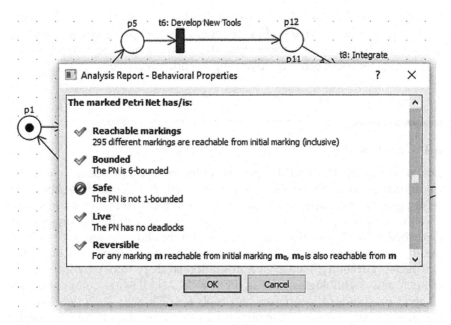

FIGURE 8.15
Behavioral analysis of the "develop system components" model.

this, the Petri nets model is adapted to the special requirements of said use case and can then be used to control the actual migration path. Starting from the initiation of the process, throughout the whole time span of the migration, the Petri net is capable of showing the current state of the process (indicated by the marking of the places), which actions need to be carried out (indicated by the transitions and its firing conditions) and which are the following steps. For this reason, the Petri net provides a model not only to plan the migration process but also to control it, when it is carried out. By integrating signals from the actual instantiated process (e.g., finishing a certain step), firing conditions for transitions can be met and the Petri nets model is therefore capable of following the actual process.

Based on this idea, a tool has been developed following the PERFoRM approach, which allows the import of a specific Petri net-based migration process model. This model is then used as the instantiation engine to supervise and control the migration process. It retrieves information from the current state of the model, which is valuable for a controlling person and displays it in a way that is easy to understand and doesn't require any Petri net knowledge. This includes all currently open steps and the according actions that need to be taken. The dependencies and the structure of the migration process are still managed by the Petri net in the background, but the person using the tool only gets filtered information about the currently important steps. In addition to displaying the current situation, the tool also provides interfaces to give feedback to the Petri net. This is used to be able to confirm that certain steps are finalized. Since Petri nets allow modeling parallel processes, it is possible that the tool is displaying various open steps at the same time. In the first version of the tool, aspects of time planning are not integrated.

8.4 Conclusions

The kernel of the PERFoRM approach is the conceptual transformation of traditional production systems into digitalized, harmonized, flexible, and reconfigurable production systems. The PERFoRM reference architecture and the various applications and tools developed within the project and described in the previous chapters represent a valid solution to decentralize the automation control in manufacturing toward increased flexibility and reconfigurability. However, a migration strategy to deploy in industry of such new technologies, that is, PERFoRM Middleware component (see Chapter 6), standard interfaces, and tools (see Chapter 7), is necessary to encourage manufacturers from migrating to next-generation industrial automation systems and guarantee the successful deployment.

A stepwise migration approach can support manufacturers to break down the path toward the vision of Industry 4.0. The five-phase PERFoRM migration process enables the identification, design, and implementation of each migration step through the Preparation, Options Investigation, Design, Implementation, and Deployment phases.

The phases are described using the Petri nets formalism, which can be simulated, validated, and controlled. The main benefit of the application of Petri nets within the control of the process is given by its clear usability and understandability by process participants. The use of Petri nets is empowered by the possibility of performing quantitative and qualitative analyses to the process.

The described migration approach has been applied to the four industrial use cases described in Chapters 10 through 13, revealing the validity and effectiveness of the conceptual approach for handling the migration toward the next generation of production systems step by step.

References

ARC Advisory Group. 2007. "Siemens Process Automation System Migration and Modernization Strategies [White Paper]" Accessed October 15, 2018. https://docplayer.net/11862216-Arc-white-paper-siemens-process-automation-system-migration-and-modernization-strategies-thought-leaders-for-manufacturing-supply-chain.html

Beserra, Patricia V., Alessandro Camara, Rafael Ximenes, Adriano B. Albuquerque, and Nabor C. Mendonca. 2012. "Cloudstep: A Step-by-Step Decision Process to Support Legacy Application Migration to the Cloud." In *2012 IEEE 6th International Workshop on the Maintenance and Evolution of Service-Oriented and Cloud-Based Systems (MESOCA): 24 September 2012, Trento, Italy,* 7–16. Piscataway, NJ: IEEE.

Cachada, Ana, Flávia Pires, José Barbosa, and Paulo Leitão. 2017. "Petri Nets Approach for Designing the Migration Process towards Industrial Cyber-Physical Production Systems." In *Proceedings IECON 2017—43rd Annual Conference of the IEEE Industrial Electronics Society: China National Convention Center, Beijing, China, 29 October–01 November, 2017,* 3492–3497. Piscataway, NJ: IEEE.

Calà, Ambra, Arndt Luder, Ana Cachada, Flávia Pires, José Barbosa, Paulo Leitão, and Michael Gepp. 2017. "Migration from Traditional towards Cyber-Physical Production Systems." In *2017 IEEE 15th International Conference on Industrial Informatics (INDIN): University of Applied Science Emden/Leer, Emden, Germany, 24–26 July 2017: Proceedings,* 1147–1152. Piscataway, NJ: IEEE.

Cetin, Semih, N. I. Altintas, Halit Oguztuzun, Ali H. Dogru, Ozgur Tufekci, and Selma Suloglu. 2007. "A Mashup-Based Strategy for Migration to Service-Oriented Computing." In Pervasive Services, IEEE International Conference on, 169–172. Piscataway, NJ: IEEE.

Colombo, Armando W., Ricardo Carelli, and Benjamin Kuchen. 1997. "A temporised Petri net approach for design, modelling and analysis of flexible production systems." *Int. J. Adv. Manuf. Technol.* 13 (3): 214–226. doi:10.1007/BF01305873.

Colombo, Armando W., Thomas Bangemann, Stamatis Karnouskos, Jerker Delsing, Petr Stluka, Robert Harrison, et al. 2014. *Industrial Cloud-Based Cyber-Physical Systems.* Cham: Springer International Publishing.

Critical Manufacturing. 2015. *MES Migration Strategies.* Accessed October 15, 2018. https://www.criticalmanufacturing.com/en/newsroom/press-releases/posts/press-releases/critical-manufacturing-offers-mes-migration-strategies-white-paper#.XVQlxUfgqUk

David, René, and Hassane Alla. 1994. "Petri nets for modeling of dynamic systems." *Automatica* 30 (2): 175–202. doi:10.1016/0005-1098(94)90024-8.

Delsing, Jerker, Fredrik Rosenqvist, Oscar Carlsson, Armando W. Colombo, and Thomas Bangemann. 2012. "Migration of Industrial Process Control Systems into Service Oriented Architecture." In *IECON 2012: 38th Annual Conference on IEEE Industrial Electronics Society; Montreal, Quebec, Canada, 25–28 October 2012; Proceedings,* 5786–5792. Piscataway, NJ: IEEE.

EU FP7 AESOP Consortium. 2010–2014. "ArchitecturE for Service-Oriented Process—Monitoring and Control (IMC-AESOP)." Accessed October 10, 2018. https://cordis.europa.eu/project/rcn/95545_de.html.

Feldmann, K., and A. W. Colombo. 1998. "Material flow and control sequence specification of flexible production systems using coloured Petri nets." *Int. J. Adv. Manuf. Technol.* 14 (10): 760–774. doi:10.1007/BF01438228.

Feldmann, K., C. Schnur, and W. Colombo. 1996. "Modularised, distributed real-time control of flexible production cells, using Petri nets." *Control Eng. Pract.* 4 (8): 1067–1078. doi:10.1016/0967-0661(96)00105-0.

Fuentes-Fernández, Rubén, Juan Pavón, and Francisco Garijo. 2012. "A model-driven process for the modernization of component-based systems." *Sci. Comput. Program.* 77 (3): 247–269. doi:10.1016/j.scico.2011.04.003.

Givehchi, O, K Landsdorf, P Simoens, and AW Colombo. 2017. "Interoperability for industrial cyber-physical systems: An approach for legacy systems.". IEEE Transactions on Industrial Informatics 13 (6), 3370–3378.

HSO. 2016. "Choosing the right ERP implementation strategy."Accessed October 15, 2018. https://www.coursehero.com/file/36003281/Choosing-The-Right-ERP-Implementation-Strategypdf/

Ionita, Anca D., Grace A. Lewis, and Marin Litoiu. 2013. *Migrating Legacy Applications: Challenges in Service Oriented Architecture and Cloud Computing Environments.* Hershey, PA: IGI Global (701 E. Chocolate Avenue Hershey Pennsylvania 17033 USA). http://services.igi-global.com/resolvedoi/resolve.aspx?doi=10.4018/978-1-4666-2488-7.

ISA Organization. 2018. "ISA95." Accessed October 15, 2018. https://www.isa.org/isa95/.

Leitão, Paulo, Armando W. Colombo, and Francisco Restivo. 2006. "A formal specification approach for holonic control systems: The ADACOR case." *Int. J. Mach. Tool. Manu.* 8 (1/2/3): 37. doi:10.1504/IJMTM.2006.008790.

Lewis, G., E. Morris, D. Smith, and L. O'Brien. 2006. "Service-Oriented Migration and Reuse Technique (SMART)." In *13th IEEE International Workshop on Software Technology and Engineering Practice, 2005: STEP 2005; 24–25 September 2005, Budapest, Hungary; Proceedings; Including Eight Workshops,* edited by Kostas Kontogiannis, 222–229. Piscataway, NJ: IEEE.

Madkan, P. 2014. "Empirical study of ERP implementation strategies-filling gaps between the success and failure of ERP." *Int. J. Inf. Comput. Technol.* 4 (6): 633–642.

Mendes, J. M., Axel Bepperling, João Pinto, Paulo Leitão, Francisco Restivo, and Armando W. Colombo. 2009. "Software Methodologies for the Engineering of Service-Oriented Industrial Automation: The Continuum Project." In *33rd Annual IEEE International Computer Software and Applications Conference, 2009: COMPSAC '09; 20–24 July 2009, Seattle, Washington, USA; Proceedings,* edited by Iqbal Ahamed and Elisa Bertino, 452–459. Piscataway, NJ: IEEE.

Open Text. 2009. "Top 10 Best Practices in Content Migration."

Suzuki, Ichiro, and Tadao Murata. 1983. "A method for stepwise refinement and abstraction of Petri nets." *J. Comput. Syst. Sci.* 27 (1): 51–76. doi:10.1016/0022-0000(83)90029-6.

Transvive. 2011 "Migration Strategies & Methodologies [White Paper]". Accessed on October 15, 2018. https://www.coursehero.com/file/30077677/Transvive-MainframeMigrationStrategy-WPpdf/

Valette, R. 1979. "Analysis of Petri nets by stepwise refinements." *J. Com. Syst. Sci.* 18 (1): 35–46. doi:10.1016/0022-0000(79)90050-3.

Wu, Bing, Deirdre Lawless, Jesus Bisbal, Ray Richardson, Jane Grimson, Vincent Wade, et al. 1997. The Butterfly Methodology. A Gateway-free Approach for Migrating Legacy Information Systems." In *Engineering of Complex Computer Systems, 3rd International Conference.* Los Alamitos: IEEE Computer Society Press. Accessed on October 15, 2018. https://www.researchgate.net/publication/3711815_The_Butterfly_Methodology_a_gateway-free_approach_for_migrating_legacy_information_systems.

Zhou, MengChu, and Frank DiCesare. 1993. *Petri Net Synthesis for Discrete Event Control of Manufacturing Systems.* Boston, MA: Springer US.

Zhou, MengChu. 1995. *Petri Nets in Flexible and Agile Automation.* Boston, MA: Springer US.

9

Test Beds

Nandini Chakravorti, Evangelia Dimanidou, Mostafizur Rahman
(The Manufacturing Technology Centre (MTC))

André Hennecke
(Technologie Initiative SmartFactory KL)

Jeffrey Wermann, Armando W. Colombo
(Institute for Informatics, Automation and Robotics (I²AR))

CONTENTS

9.1 Overview of Test Beds and Their Use within PERFoRM

The validation of the overall architectural principles behind the Production harmonizEd Reconfiguration of Flexible Robots and Machinery (PEFoRM) approach was conducted by applying the architectural concepts to four diverse use cases: (1) manufacturing of tailored compressors trains for oil and gas applications, (2) assembly of low-cost full electric vehicles with high variants and high quality on low-budget assembly lines, (3) production of aerospace components within high variants, and (4) production of consumer white goods.

Prior to the actual deployment of the new developed technologies at the end users' industrial premises, the basic principles of the PERFoRM system architecture and technologies were validated within the industry-like environments of the Test Beds. PERFoRM's Test Beds were set at the Manufacturing Technology Centre (MTC 2018) and SmartFactory (SmartFactory 2018), with both centers equipped with diverse production resources that were made available for testing the developed concepts.

The MTC is a UK high value manufacturing Catapult centre (Innovate UK 2018), created to bridge the gap between fundamental research and industrial applications by providing a real manufacturing environment where technology can be matured and de-risked with the support of skilled and experienced engineers. The MTC has four key technology themes—Digital Engineering, Advanced Production Systems, Component Manufacturing Technologies, and Intelligent Automation. The MTC has a membership of more than 100 companies in the manufacturing, technology, and services domain, providing expertise on migrating existing manufacturing systems to new technology readiness levels.

SmartFactory (2018) is a non-profit small- to medium-sized enterprise (SME) located in Germany that implemented and demonstrated the innovation potential of PERFoRM through its manufacturer-independent research and demonstration platform. The last is operated by the SmartFactory-Network, consisting of more than 50 members. An important element of SmartFactory's mission is to move innovations as quickly as possible from the lab into the real factories. SmartFactory builds and operates various Information and Communications Technology (ICT)-based production lines, which contributed in the investigation and prototypical implementation of the PERFoRM migration approach and in the multiplication of the solutions both immediately through the SmartFactory-Network and through its extensive Public Relations (PR) activities.

Testing activities involved assets and supporting infrastructure available at each test bed. A detailed list of hardware assets (such as robots, controllers, machines, and databases) that are available at the test beds (MTC and

SmartFactory) and four pre-Test Beds (e.g., Loccioni (AEA SRL 2018), TUBS (Technical University Braunschweig 2018), IPB (Polytechnic Institute of Braganza, Portugal 2018), and I²AR (Institute for Industrial Informatics, Automation and Robotics, University of Applied Sciences Emden/Leer, Germany 2018)) were created in order to effectively distribute the innovation activities among the involved actors and avoid duplication of testing efforts.

The work performed on the test beds focuses on the verification and validation of the technological developments to confirm that user requirements are met and to de-risk solutions before they are implemented in operational environments. The demonstrators were used to test and optimize the developed solutions and to provide realistic and valuable information for the development of migration strategies.

Concurrent development and incremental validation of the digitalized reconfigurable and self-adaptive systems at industrial scale (at MTC and SmartFactory) allowed exploration of PERFoRM's innovation work and rapid iterations to guide and de-risk translation to industrial deployments with minimum disruptions.

9.2 Whirlpool Use Case Objective and Architecture

Whirpool's motivation was to develop solutions to minimize changeover effort and allow fast integration of automation systems into the wider company ICT infrastructure (Whirlpool Corporation 2018). Additionally, the visibility of real-time production system status for decision makers was also identified as a key driver. The overall objectives of this use case were (1) monitoring of Key Performance Indicators (KPIs) related to detection of performance degradation and (2) fast reconfiguration of the path of a robot equipped with a probe for the detection of microwave leaks coming from a microwave oven.

The fast reconfiguration of a robot was selected for validation and demonstration within MTC's preindustrial environment. The robot, a Universal Robot 10 (UR10), was used to perform an electromagnetic compliance (EMC) leakage test based on special markers on the Computer-aided Design (CAD) file of the microwave. The system consists of a UR10 robot, equipped with a microwave probe to enable the detection of microwave leakages. The scenario was mapped to the overall PERFoRM architectural concepts as seen in Figure 9.1.

The robot moves the probe following a predefined path around the microwave oven. The CAD file contains the information for the probe path and is stored in a file repository. This CAD file is automatically transferred to the Adapter via the Middleware. The markers in the CAD file are translated to a UR path file via the adapter. Finally, the adapter automatically transfers the UR path file to the UR10 robot for conducting the leakage testing.

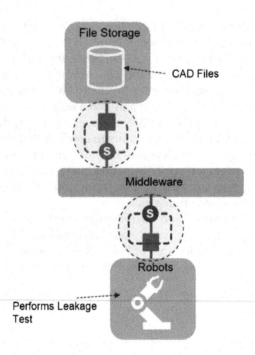

FIGURE 9.1
Whirlpool leakage test architecture. (From MTC 2017.)

If leakages are detected, the oven is sent for repair. The role of the adapter is to automatically generate the robot commands starting from a path drawn with a CAD tool and send them to the robot. This permits the product designers to directly draw the most appropriate path for detecting possible microwave leakages outside the oven. Moreover, it promotes no stoppages to the production line when a new model of oven is released to production as it enables an automated replacement of the path program.

9.2.1 Demonstrator Setup

The first iteration of the test bed was set up at the MTC, Coventry, UK. The implemented architecture is as shown in Figure 9.2, which includes A UR10, equipped with a 3-D printed probe for testing the leakages at the outside of a microwave oven, the PERFoRM Middleware (Apache ServiceMix) following the specifications reported in Gosewehr et al. (2017), a file transfer protocol (FTP) server for storing the CAD files, and the UR Adapter that translates a particular CAD file to a URScript and transfers it to the robot.

For the purpose of the demonstrator, four virtual machines (VMs) were set up to replicate the decoupling between the individual systems. The VMs also allow a user to directly copy the VM image to individual host machines, which can result in four different host PCs running the File Server, Operator

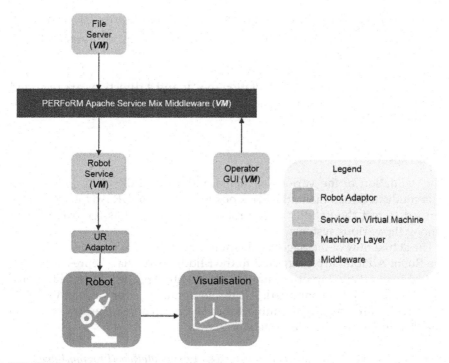

FIGURE 9.2
Demonstration of the Leakage test of microwaves using the PERFoRM compliant Middleware (Apache ServiceMix). (From Chakravorti et al. 2017.)

Graphical User Interface (GUI), Robot Service, and the Middleware, respectively (see Figure 9.2). The functions of each of the VMs are listed below:

- File Server VM: Hosts the CAD files for performing the robot leak test.
- Operator GUI VM: Reads the list of CAD files from the File Server and displays them such that an operator can select a particular file.
- Robot Service VM: Is used to pull a particular CAD file from the File Server and push it to a predetermined folder that the UR Adapter can access.
- Middleware VM: Comprises an Apache ServiceMix.

The sequence of operations for this demonstrator can be seen below:

- An operator selects a CAD file from the list displayed on the Operator GUI. This results in the Operator GUI Service sending a trigger to the Robot Service, which then calls the FTP service requesting the file selected by the operator.
- The selected CAD file from the File Server VM is downloaded, communicating via the Middleware, to a folder that the Robot Service can access.

- Once the Robot Service gets the file, it pushes this file to the UR Adapter.
- The Adapter translates the CAD file to a URScript and pushes the script to the robot.

The linear movements and the rotation in x, y, and z directions are recorded using a test script. The actual and the planned movements are being visualized in a Matlab graphic.

9.2.2 Results

The validation of the various elements within the PERFoRM architecture was conducted incrementally. Black box testing of the UR Adapter was performed to test the validity of the robot script, as well as the overall flow among the various systems.

The actual flow of messages is illustrated in Figure 9.3. The File Server and the Robot Adapter are connected to the Middleware, via a representational state transfer (REST) interface implemented with Apache CXF. Additionally, the Operator GUI is connected. All the components offer various Web services, which are used to communicate between each other. The flow sequence depicted in Figure 9.3 is described as follows:

- Step 1: The Operator GUI invokes the *updateRobotProgramTable( )* service. This service is used to display the list of available robot programs and their status (e.g., "buildmicrowave" has a status indicating that it is running, as seen in Figure 9.3).
- Step 2: The Middleware uses the *getCurrExecProgID( )* service provided by the Robot Service to get the current program, which is in execution.
- Step 3: The current executing program is available to the Middleware.

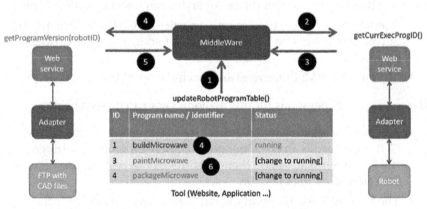

FIGURE 9.3
Flow of messages in the Whirlpool Demonstrator. (From Chakravorti et al. 2017.)

- Step 4: The Middleware uses the *getProgramVersion( )* service provided by the File Server to retrieve the programs stored on the File Server.

- Step 5: The programs stored on the File Server are now available to the Middleware.

In a similar way, the GUI can be used to enable the operator to select a program from the list. Using the updateRobotProgram( ) function the selected CAD file is being sent through the Middleware to a specific folder. The UR adapter, which has the ability of monitoring the specific folder, consumes the file as soon as it detects it.

It is to be noted that this demonstrator can be extended to include further production assets. Using a setup of four different VMs, the files incoming from the host via C:\temp\ftp were successfully transmitted via FTP to the outgoing Robot Service, which writes the output file to C:\temp\robot. The adapter was then able to consume the file to drive the UR10. This demonstrates vertical integration from the FTP file server, via the Middleware, down to the Robot Service and the Adapter, which finally transmits the new program to the actual robot.

A program is also written on the robot controller to record the actual positions reached by the robot arm. A comparison between the path specified in the CAD file and the actual recorded positions can be seen in Figure 9.4. The comparison error is currently ±12 mm. This calculated error is much higher than the actually observed during the trials; a possible reason for this large error may be because the reference path does not account for the robot's acceleration. The reference path assumes that the robot moves from point to point at a constant speed,

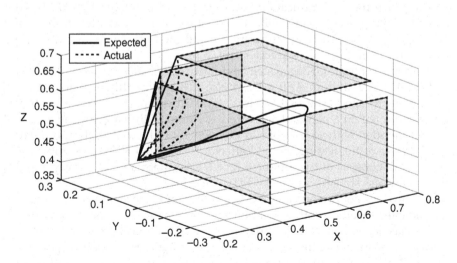

FIGURE 9.4
Comparison of path in CAD file with actual movements by robotic arm. (From Chakravorti et al. 2017.)

whereas, in practice, the arm will accelerate to achieve the commanded speeds. To get a more accurate figure, the robot's speed profile will need to be known and accounted for in each move. Another approach could be to ignore the temporal characteristics of the robot and calculate a spatial-only error. Further work will need to be conducted to determine if this error can be minimized.

Moreover, in order to eliminate the danger of the robot crashing into the oven while moving from one point to the next, the robot returns to a predefined position every time the robot scans a side of the oven. This is visualized in Figure 9.4 at the far left corner of the plot.

9.3 GKN Use Case Objective and Architecture

GKN aims to implement a Micro-Production Flow Cell that is able to maximize the flexibility and agility of the production processes, reduce lead times, and improve resource utilization (GKN Group 2018). The innovative reconfigurable robotic cell designed on the basis of the PERFoRM approach was planned to be a heterogenous system, where Open Platform Communications Unified Architecture (OPC-UA) standard is used as an integration layer that enables the communication between the asset and higher Enterprise level.

9.3.1 OPC-UA Demonstration and Validation Approach

The main goal of GKN's preindustrial demonstrator developed at the MTC was to prove the communication between the cell robot, an ABB 6700, and any other external entities via OPC-UA. Custom OPC-UA adapters have been developed to allow the communication using the Middleware architecture and the nodes in the GKN's architecture.

In order to replicate GKN's system architecture, a *Small Component Assembly Cell (SCALP cell)* and an *ABB robot (model: IRB 140)* readily available at the MTC were selected. The IRB 140 is a standard compact six-axis industrial robot, mounted on a pedestal and safely enclosed within a guarding area. It can be provided with a series of end-effectors for industrial applications. For the purpose of the testing, no end effector was bolted at its extremity. The SCALP cell consists of two underslung KUKA robots: a servo conveyor and several process stations and tools (MTC 2017).

As ABB does not provide an OPC-UA toolkit for these legacy robots, a custom adapter together with a solution for downloading the NC/Robot programs was developed to enable and prove the communication with the robot. Furthermore, for one of the planned processes (roughness inspection process) in the GKN use case, an adapter is required for substituting the wired serial communication of the Mitutoyo roughness sensor. The integration of the sensor adapter within the PERFoRM framework was also validated.

9.3.1.1 OPC-UA Demonstrator Setup

A physical demonstrator for the validation of some of the main features of the reconfigurable flexible cell that is implemented at GKN's site was deployed at the MTC. The main challenge of this demonstrator was to test the OPC-UA connectivity infrastructure that is used as an integration layer between the IT Level and the Asset Level. In order to further extend the current setup, the XETICS LEAN environment (see XETICS GmbH 2018) has been selected to host the robot programs and replicate the HI-FIT system, used by GKN. The architecture of the demonstrator is illustrated in Figure 9.5 and involves the following systems:

- DNC/Robot Programs, stored in the HI-FIT repository. For the testing purposes, a File server was used.
- OPC-UA communication protocol to transfer data between the systems.
- Robot controller programs.
- ABB IRB 140 robot.
- SCALP cell.
- XETICS Manufacturing Execution System (MES).

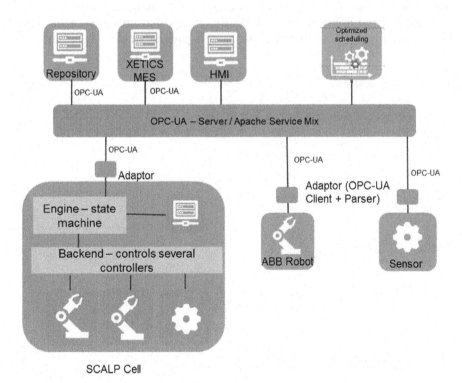

FIGURE 9.5
GKN's physical demonstrator architecture at MTC (2017).

Custom OPC servers and clients were developed using the .NET programming environment (C#) for communicating with the IRB 140 robot and the MTC's SCALP cell.

The ABB robot server exposes data, such as the robot's current operating mode, mastership and IO signal values, via the OPC-UA protocol. The communication to the ABB robot is handled using ABB's PC Interface, which is designed to allow data acquisition from a robot controller over a network, and the ABB PC SDK Interface to gain mastership of the robot for remote control. On start-up, the software scans the connected network for compatible controllers and then creates individual PC and PC SDK Interface instances for each of them. Each instance creates a representation of the controller data using OPC-UA nodes, which can be read by an OPC-UA client. The PC Interface is then used to poll the robot controller for data, which is used to update the OPC-UA nodes. The PC SDK Interface is used only for sending commands. For the purpose of this work, as well as for safety reasons, the connection was deliberately restricted to this particular IRB 140; nevertheless, it can easily be extended to any networked ABB.

The OPC-UA server (Figure 9.6) of the SCALP cell allows internal cell data to be read externally by an OPC-UA client (Figure 9.7). This data includes information such as the current program, safety state, and tool data. The SCALP cell publishes several C# WCF services that enable control of the cell externally over a network. Using a set of previously developed wrapper

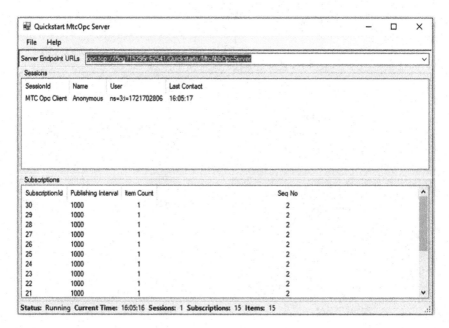

FIGURE 9.6
OPC Server Adapter. (From MTC 2017.)

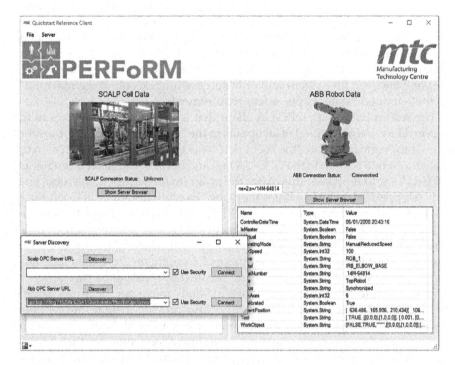

FIGURE 9.7
Client application for monitoring server data. (From MTC 2017.)

libraries, communication and data exchange with the SCALP cell is scripted using the WCF services. The adapter handles the SCALP cell data and updates the OPC-UA data representation.

Both OPC-UA servers (IRB 140's and SCALP cell's) handle variables writing requests via designated event handlers (for writable variables). However, taking into consideration systems' safety, this capability has been purposefully limited to only certain preselected systems data.

The client hosted in a Windows machine provides a UI for discovering, connecting, and browsing the ABB and SCALP cell OPC-UA data as well as a dialog box for issuing Write requests.

Moreover, an MES developed by XETICS was also used for this proof of concept. The XETICS LEAN communicates with the SCALP cell and ABB robot using the aforementioned OPC-UA servers via an additional XETICS module called Equipment Integration (EI). The EI module is responsible for reading the robot variables (such as robot position, robot speed, robot current tool number, robot pneumatic pressure, robot state, current work object, axis 1-6 calibration status, and execution status) and writing the variable Program to run.

The system is designed to store the data acquired from the robots in a local data repository (InfluxDB) and make them become available to the user via the XETICS Interface. Additionally, the user can execute a robot program from the XETICS UI by performing a "move in" ("move in" is a term used by XETICS to define when an order is being processed on a machine, thus making the machine busy) on a preconfigured order that writes the respective robot program execution code to the "CommandRun" tag in the OPC-UA server. However, the system will only successfully execute if the machine is in the "stopped" state, as per safety requirements.

The ABB robot has an OPC-UA client that is subscribed to the tags in the server. Upon being notified of an update in the "CommandRun" tag, the robot goes into "running" state (the robot performs the operation respective to the execution code written to the OPC-UA tag) and updates the "ExecutionStatus" OPC-UA tag to represent this. After the program has finished, the robot goes into "stopped" state and updates the same "ExecutionStatus" OPC-UA tag as before to represent the status change. XETICS LEAN reads this and performs a "move out" (move out is a term used by XETICS to define when an order has finished being processed by a machine, thus making the machine free for another order) for the specified piece of equipment. Once the machine is in "stopped" state, XETICS can trigger the next program to be run.

In addition to the automated "move in" and "move out" features driving the functional operations, these features also provide insight into KPI performance. For example, the time between "move in" and "move out" can be measured as "actual operational time." This operational time can then be compared against the likes of planned operational time to check actuals against plans. It can also be compared against machine available time, to get information on machine utilization. The following KPIs are measured by XETICS:

- Quality (Calculated using scrapped parts and good parts data).
- Uptime (Calculated using standby time and productive time data).
- Downtime (Calculated using maintenance time, ramp up time, and repair time data).
- Availability (Calculated using uptime and downtime data).
- Performance (Calculated using productive time and standby time data).
- Overall Equipment Effectiveness (OEE) (Calculated using the availability, quality, and performance data calculated above).

9.3.1.2 Results

The OPC-UA connectivity infrastructure was tested incrementally to proof the concept for the GKN use case. ABB does not provide an OPC-UA toolkit that would enable access to legacy robot's controller. However, via the PC Interface option, which comprises a robot side communication plug-in and a .NET Software Development Kit (SDK), communication between the controller and external software was achieved. The plug-in along with .NET Kit forms the PCSDK option, which exposes a series of commands and controller data to a connected application. The application can then browse the network for available controllers, connect to them and read data. By default, only read access is granted to a robot. In order to gain write access, the robot must be in automatic mode and slaved to the controlling application. The ABB OPC-UA server uses PCSDK to find available controllers on the network, connects to them, and then wraps the exposed data into the correct format for adding to an OPC-UA server. This allows the robot's internal data structures to be exposed externally by an OPC-UA client.

For the purpose of testing, a standard file repository located on a common network was used to represent the HI-FIT repository, store the robot programs, and was used to verify the remote transferring of programs to both the SCALP cell and ABB robot via OPC-UA. Moreover, the Mitutoyo sensor was integrated with the Apache ServiceMix, which acts as an OPC-UA proxy server, replicating the OPC-UA server and therefore allows a seamless transition from the old system to the new (i.e., wired to unwired) that implements PERFoRM's architecture. The OPC-UA service that was developed to enable the communication between the sensor and the Apache ServiceMix allows the decoupling of the server from the client, giving the flexibility of supporting various communication protocols, based on user's needs. Detailed testing of the system has been conducted to validate that the sensor readings were correctly and consistently available to OPC-UA client through the PERFoRM's Middleware, specified in Gosewehr et al. 2017. Finally, both, the SCALP cell and the IRB 6700 ABB robot were successfully integrated with XETICS LEAN. The data retrieved from the robots are plotted on trend graphs, demonstrating clearly the integration of low-level controls to manufacturing operation control systems, such as MES which can be used for the purposes of planning and scheduling.

9.3.2 Demonstration of the PERFoRM Architecture in a Reconfigurable and Modular Environment

One of the key results of applying the PERFoRM approach is the definition, specification, development, and prototype implementation of a system architecture and concepts enabling highly flexible production systems. To validate this architecture and de-risk the technologies before the implementation into a real production systems, a modular and reconfigurable demonstrator was setup at SmartFactory^KL—a mini flexible cell.

The mini flexible cell was designed to replicate the flexible robot cell designed by GKN, but it also covers parts of the other use cases. It is implementing the concept architecture and its key elements, in particular the technology adapter, the Middleware, the harmonization of different communication protocols, and the integration of advanced tools. Those solutions will be partly adapted, for example, to visualize the state or KPIs with data produced by the demonstration environment and not with the data from the industrial use cases. But it also acts as a real test study, for example, by using a Programmable Logic Controller (PLC) used by GKN and the PLC-adapter to show the integration of legacy PLCs to OPC-UA or by using the reconfiguration tool for the reconfiguration ability of the cell, which will also be used at GKN.

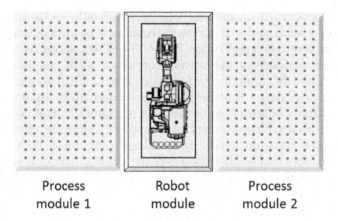

Process	Robot	Process
module 1	module	module 2

FIGURE 9.8
Basic concept of the mini flexible cell. (From EU HORIZON 2020 FoF PERFoRM 2015-2018.)

Figure 9.8 depicts the basic concept of the demonstrator which is based on the use case idea. It includes a central robot module hosting a collaborative robot and interchangeable process modules which can be placed on any side of the robot module.

9.3.2.1 Design and Demonstrator Setup

Each process module is designed to host active or passive, hot plug-in capable subsystems. Currently an automated press station and an automated turntable, acting as a storage system, are available which were designed as Cyber-Physical Systems (CPSs). The subsystems work together with the robot during the process and are easy exchangeable according to the Plug-and-Produce principle, in order to show the full reconfigurability of the cell. Figure 9.9 shows the design and the first setup of the demonstrator.

FIGURE 9.9
Design and first setup of the demonstrator. (From Hennecke and Ruskwoski 2018.)

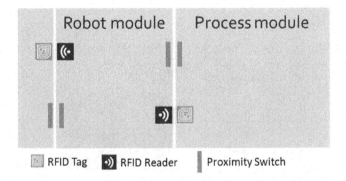

FIGURE 9.10
Neighborhood-based topology detection concept. (From Hennecke and Ruskwoski 2018.)

To detect the layout of a cell, a detection mechanism is implemented between the central robot module and the process module. This Neighborhood-based topology detection concept is depicted in Figure 9.10. It consists of a proximity switch to detect if a new process module appears and an RFID system for the identification of a process module by using the unique identifier which is contained on the RFID tag.

9.3.2.2 Instantiation of the PERFoRM Concept Architecture

PERFoRM recommends a concept system architecture, depicted in Figure 9.11 (right). This demonstrator was set up to validate that the concept architecture enables reconfigurable systems; therefore, all of the key elements were implemented as described in the left picture of Figure 9.11. Table 9.1 gives an overview of the concept architecture elements and how or why it was implemented in the demonstration architecture.

TABLE 9.1

Mapping between the PERFoRM Architecture Elements and the Implemented and Demonstrated Elements

ID	Concept Architecture Element	Implementation and Demonstration
1	Adapter to legacy machine	OPC-UA Adapter to a legacy PLC (S7-300)
2	Adapter to legacy robot	OPC-UA Adapter to a collaborative robot controller (UR5)
3+4	Communication protocols: REST, message queuing telemetry transport (MQTT), and OPC-UA	Implemented by using REST and MQTT for the communication between the Middleware and the tools and OPC-UA as the connectivity standard to machines and robots
5	Middleware	Apache Camel as a routing engine between different protocol and data storage endpoints
6	Advanced tools integration	Implemented and demonstrated by a reconfiguration tool and a mobile order client
7	Connection to a database	Connection from the Middleware to a No-structured query language (SQL) database

Source: Hennecke and Ruskwoski (2018).

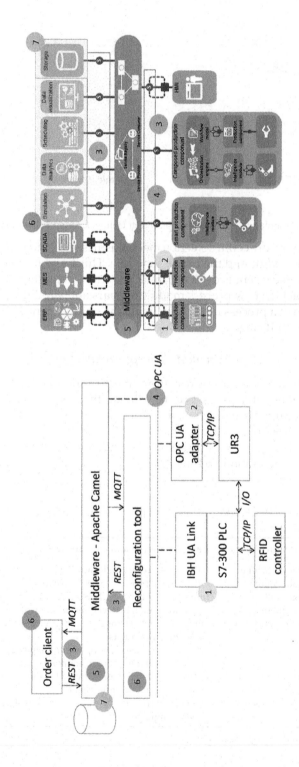

FIGURE 9.11

Instantiated architecture and elements (left) compared to the concept architecture and its elements (right). (From Hennecke and Ruskwoski 2018.)

9.3.2.3 Implementation and Connectivity

To enable the reconfiguration on cell level, the modules and the robot are encapsulated and integrated with OPC-UA, as illustratively shown in Figure 9.12. Each module and robot provides an OPC-UA server representing a uniform information model and providing a low-level interface to access the controller. The types of the model are compatible with IEC 61131-3 and current high-level programming languages (C#, C++). Only integer data types are used for numeric data.

Three information models were defined: one for the robot module, one for the robot itself, and one for the process modules.

The realization of this architecture was done by implementing the software components specifically for this demonstrator and for demonstration and architecture validation purpose, namely:

1. **OPC-UA Adapter** to a Universal Robot and the PLCs, providing the OPC-UA information model and connecting to the low-level control systems.

2. **Reconfiguration Tool** to discover, monitor, and manage the modules and the cell layout as well as the process coordination mainly with the OPC-UA information from the modules.

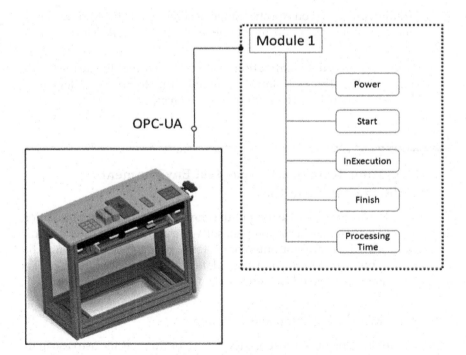

FIGURE 9.12
Module integration with OPC-UA. (From SmartFactory-KL 2017.)

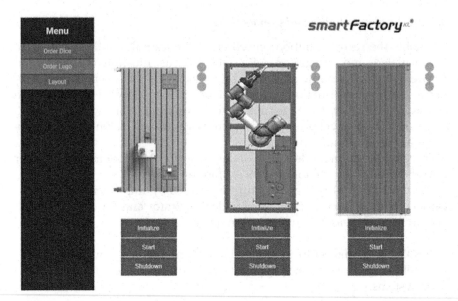

FIGURE 9.13
Visualization of the cell layout in the monitoring tool.

3. **Middleware,** based on Apache Camel to demonstrate the Middleware aspect of this project and the connectivity to OPC-UA, MQTT, and databases.

4. **Order Client and Visualization** is a web-based tool to start orders and visualize the cell status by connecting to the Middleware. Figure 9.13 shows a screenshot of the cell layout.

9.4 Integration Tests in a Virtual Test Environment for the Siemens Use-Case

The use case from Siemens, dealing with schedule optimization, simulation, and predictive analytics, has a complex software architecture involving different services which are not only connected with each other through the Middleware, but also to the shop floor database and legacy software. The logical overview is depicted in Figure 9.14.

9.4.1 Test Environment Setup and Test Steps

To test the integration of all the tools developed for this specific use case in the PERFoRM architecture, a virtual test environment was set up by using Virtual Box and a local server infrastructure. This set up is depicted in Figure 9.15.

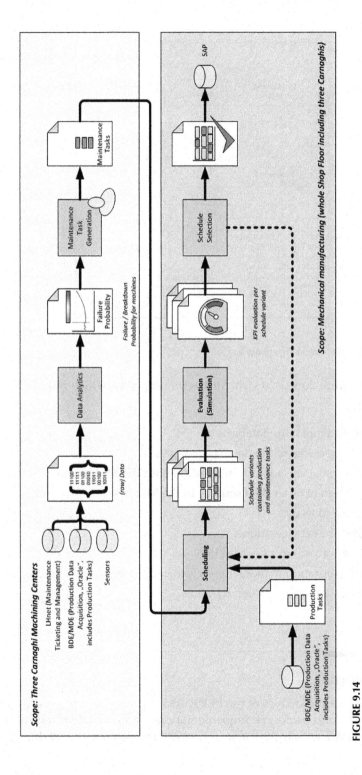

FIGURE 9.14
Siemens use-case—Logic workflow overview. (From SmartFactory-KL 2017.)

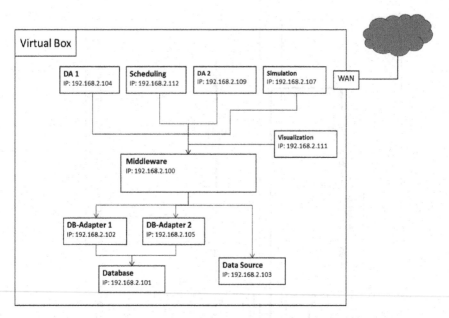

FIGURE 9.15
Siemens Use-Case—Workflow overview.

The tests were done in a structure bottom-up way, which follows the following steps:

1. Setup of a mock-up database.
2. Test the services of the database adapter.
3. Test the database adapter integration with the Middleware.
4. Integration of other data sources to the Middleware.
5. Integrate tools by:
   a. Test the services alone.
   b. Integration with the Middleware.
   c. Collaboration to mock-up service endpoints.
   d. Collaboration between the tools.

9.5 Conclusion

One of the main objectives of the PERFoRM approach was the specification, development, and prototype implementation of an innovative control, automation, and ICT architecture able to support a new generation of digitalized, harmonized, and reconfigurable manufacturing systems working on the basis of plug-and-produce concepts, thus enabling an agile production of smaller lot sizes, more customized products, shorter lead times, and shorter time to market. The Test Beds were responsible for de-risking the technologies developed for supporting the proposed architecture, before they were applied to four different industrial manufacturing production lines. The validation and testing results obtained from the test beds have been used to optimize the deployment strategy, minimize any potential risks, and develop the migration plans for the four use cases, enabling the successful deployment of the PERFoRM approach.

References

AEA SRL. 2018. "Loccioni Group." Accessed October 14, 2018. https://www.loccioni.com/en/.

Chakravorti, N., E. Dimanidou, G. Angione, J. Wermann, and F. Gosewehr. 2017. *Validation of PERFoRM Reference Architecture Demonstrating an Automatic Robot Reconfiguration Application.* Emden, Germany: IEEE INDIN 2017.

EU HORIZON 2020 FoF PERFoRM. 2015-2018. "Deliverable D10.1: GKN Use Case Goals, Requirements and KPIs—Specification of Applications, Functions and Requirements for the Micro-Flow Cell." Unpublished manuscript, last modified October 10, 2018. http://www.horizon2020-perform.eu/index.php?action=documents.

GKN Group. 2018. "GKN Aerospace Sweden." Accessed October 14, 2018. http://www.gkngroup.com/GKNSweden/Pages/default.aspx.

Gosewehr, Frederik, Jeffrey Wermann, Waldemar Borsych, and Armando W. Colombo. 2017. "Specification and Design of an Industrial Manufacturing Middleware." In *2017 IEEE 15th International Conference on Industrial Informatics (INDIN): University of Applied Science Emden/Leer, Emden, Germany, 24-26 July 2017: Proceedings*, pp. 1160–1166. Piscataway, NJ: IEEE.

Hennecke, André, and Martin Ruskwoski. 2018. "Design of a Flexible Robot Cell Demonstrator Based on CPPS Concepts and Technologies." In Proceedings of the 1st IEEE Industrial Cyber-Physical Systems (ICPS) Conference, DOI: 10.1109/ICPHYS.2018.8390762. Accessed on October 10, 2018. https://ieeexplore.ieee.org/document/8390762

Innovate UK, Catapult. 2018. "The Catapult Program." Accessed October 14, 2018. https://catapult.org.uk/.

Institute for Industrial Informatics, Automation and Robotics, University of Applied Sciences Emden/Leer, Germany. 2018. "I2AR." Accessed October 15, 2018. http://oldweb.hs-emden-leer.de/forschung-transfer/institute/i2ar.html.

MTC. 2017. EU FoF PERFoRM Project "Self-adaptive machines demonstrator documentation and results." Accessed on October 10, 2018. http://www.horizon2020-perform.eu/

MTC, UK. 2018. "Manufacturing Technology Centre, UK." Accessed October 11, 2018. http://www.the-mtc.org/.

Polytechnic Institute of Braganza, Portugal. 2018. "IPB." Accessed October 14, 2018. http://portal3.ipb.pt/index.php/en/ipben/home.

SmartFactory. 2018. "SmartFactory[KL]." Accessed October 15, 2018. https://smartfactory.de/.

SmartFactory-KL. 2017. EU FoF PERFoRM Project "Self-adaptive highly modular and flexible assembly demonstrator design and set-up." Accessed on October 10, 2018. http://www.horizon2020-perform.eu/.

Technical University Braunschweig, Germany. 2018. "Chair of Sustainable Manufacturing and Life Cycle Engineering." Accessed October 14, 2018. https://www.tu-braunschweig.de/.

Whirlpool Corporation. 2018. "Whirlpool Europe Srl." Accessed October 15, 2018. http://www.whirlpoolcorp.com/.

XETICS GmbH, Germany. 2018. "XETICS LEAN Smart Factory." Accessed October 15, 2018. https://www.xetics.com/.

10

Use Case: Compressors

Nils Weinert

(Siemens AG)

CONTENTS

10.1 Introduction

The Duisburg factory is the headquarters of the Business Unit industrial compressor manufacturing (Siemens "Industrial Compressor Manufacturing" 2018). The factory manufactures tailored compressor trains for oil and gas applications like in the chemical and petrochemical industries, in refineries, air separation plants, or in blast furnaces. The product lines range from standardized single compressors of different physical types (single-shaft compressors or integrally geared compressors) to customized latched and geared as well as fully motorized string solutions. The factory covers the production of small batches (~30 pieces) down to one-of-a-kind manufacturing. It also operates a so-called "Mega Test Center" to conduct final acceptance tests. These cover string tests for large cracked gas compressors for mega olefin plants and large compressor strings for the customer's businesses.

The central manufacturing processes in the Duisburg factory for the compressor manufacturing are machining, heat treatment, balancing, assembly, and

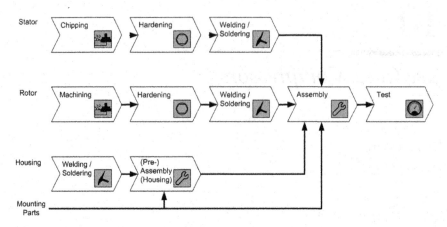

FIGURE 10.1
Principle process chain.

painting. These processes are required for all compressors—independent of their type. They can be differentiated between the two main components of the compressor, the rotor and the stator, forming the overall product together with the housing and additional mounting parts. Figure 10.1 depicts the principle process chain the manufacturing process of a compressor is running through.

10.2 Motivation

In the long term, the Siemens Compressor Factory at Duisburg is going to holistically implement the Digital Factory Paradigm. Following the general definition associated to this paradigm, this means that all necessary steps for product realization—from early planning through development to manufacturing and use of products, as well as for supportive processes like procurement, sales, etc.—are closely interconnected using information and communication technologies. Moreover, the factory is intended to achieve a strong, dedicated ability for self-adaptation, always based on the situational conditions found at a certain time. The implementation of the Production harmonizEd Reconfiguration of Flexible Robots and Machinery (PERFoRM) Demonstrator, therefore, has to be understood as one of the multiple ongoing approaches in an already transforming factory.

In the context of this Use Case selected for the validation of the PERFoRM approach, this integration and automatic adaption is fostered by the integration of the PERFoRM Middleware on the one hand, and on the other hand, by starting the implementation of a system to prevent machine breakdowns by moving to a predictive instead of a reactive and time-based maintenance

approach. Once fully established, a complete integration of the maintenance system with the Enterprise-Resource-Planning (ERP)/Manufacturing
Execution System MES (SAP) is intended, providing automated maintenance task definition based on automated data analysis of machine health
data. Therefore, although currently mainly addressing machine availability and better scheduling of maintenance tasks, the implementation of the
PERFoRM Use Case represents a first step toward the long-term goal of the
digital factory.

The Siemens Use Case aims on improving the availability and reliability
of production machinery, using the example of three comparable machines
at the Duisburg Compressor Factory (Cala et al. 2016). A direct consequence
is seen in increasing the production flexibility with more available production capacities at a time and a higher accuracy in realizing production plans
with lesser unplanned events such as machine downtimes after failures.
By preventing unplanned downtimes through an improved monitoring of
the actual condition of the production equipment, better planning of maintenance activities necessary to prevent breakdowns and repair is possible.
Further, respective measures shall be planned and executed with respect
to the overall manufacturing schedule in order to find suitable slots with
minimal negative impact on the manufacturing execution. Summarized, the
major goals to be met by applying the PERFoRM approach in this use case
are as follows:

- Raised machine availability through improved condition knowledge and maintenance activities.
- Better planning and scheduling of necessary downtimes of machinery for maintenance in order to reduce negative impacts on production figures (e.g., completion dates, downtimes, schedule deviations).

10.2.1 Intended Workflow

As introduced in the above sections, the intention of the Siemens Use
Case (Figure 10.2) is to improve the flexibility of the overall production by

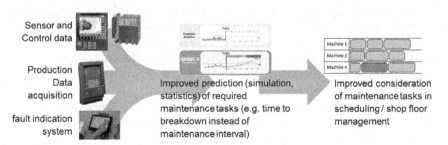

FIGURE 10.2
Use case approach overview.

raising the machine availability through better maintenance. At the same time, the maintenance tasks to be conducted shall be better scheduled with the actual manufacturing tasks in order to minimize negative effects of machines undergoing maintenance and not being available for production meanwhile.

With the system installed for the use case, several tools and services developed following the PERFoRM approach are combined to form an integrated solution supporting these targets. In the workflow created (see Figure 10.3), data available through the legacy and newly installed systems—alarm messages from the machine controls, failure tickets from the maintenance ticketing system, and sensor data from dedicated sensors applied to the machines—is analyzed in order to gain knowledge on the actual machine condition. This analysis will be conducted in parallel to the manufacturing running, thus allowing to detect condition changes over time. As a second step, the results of the Data Analysis are visualized for maintenance personnel, who will then be able to define maintenance tasks required to maintain the availability of machines for production.

As depicted in Figure 10.3, these first two steps are conducted for the three machining centers considered. Thus, the intermediate results are maintenance tasks defined with respect to the current condition of each machine. The tasks are not yet scheduled but are defined as tasks comparable to the actual production tasks, which means in general they are defined as consuming a certain amount of time on a specific resource (Machine). Additional information is an earliest (e.g., for spare parts to be available) or a latest date (e.g., after which the breakdown probability becomes too high).

In the third step, scheduling is conducted for the primary production tasks, together with the maintenance tasks. The scheduling is done for the whole manufacturing area in order to consider bottlenecks, for example, caused by delays through maintenance slots. At each time, several schedules are generated, differing, for example, in the weighting of the scheduling targets (e.g., maintenance as early and as late as possible, highest accuracy of meeting delivery date, earliness, tardiness). The defined schedules are then fed to the production simulation step. In principle, the simulation step is used to consider unforeseen events in the overall production—delays, breakdowns of (other) machinery, etc.—and their influence on the suitability of each schedule. As a result, for each schedule a set of indicators is calculated based on the simulation, thereby assisting the planning and maintenance personnel in selecting the schedule to be implemented. This is the last step of the PERFoRM Use Case System.

For the implementation within the PERFoRM Use Case, the maintenance task definition and the selection of the most suitable schedule are done manually. The data analytics solutions are intended to analyze trends in the data available in order to indicate changing machine conditions.

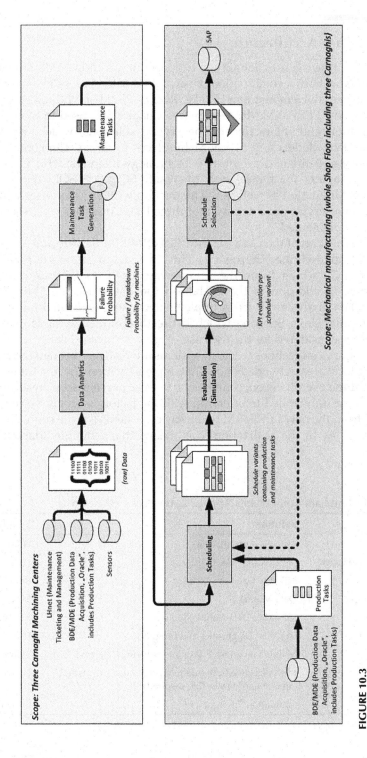

FIGURE 10.3
Intended workflow of the PERFoRM system implemented by the use case.

10.3 Use Case Architecture

For implementing the functionality to support the workflow as described in the previous section, a combination of several building blocks is integrated to form the overall Demonstrator (EU HORIZON 2020 FoF PERFoRM 2016).

As a basis, the PERFoRM Middleware (see Chapter 6) is used to ensure the communication and interaction of the several solutions, which uses the PERFoRM Data Models and Standard Interfaces to do so. The connection to the legacy systems is done via two Technology adapters, which are the Database Adapters for Oracle and Microsoft SQL (MS-SQL) databases. Scheduling, Simulation as well as the interaction with a system user will be present as one service per category, and three Data Analytics tools/services have been implemented.

For the realization of the use case, the PERFoRM Middleware and Services are hosted on dedicated hardware. This includes a standard Persoanl Computer (PC), which hosts the Middleware itself, as well as most of the services. In addition to the PC-hosted tools/services, the "Min-Max-Toolbox" Data analysis service is implemented using dedicated hardware (Microcomputer Units, Sensors), which is directly mounted in the cabinets of the machines considered for the use case.

The necessary connections to legacy systems are realized through the manufacturing local area network. For the first implementation, the first transition step toward the overall vision, connections to legacy systems are "read-only" for preventing unintended interference with the factory systems. Necessary data flow from the new PERFoRM to the legacy systems ("write connection") is set manually. In the realization, this means that schedule adaptations

TABLE 10.1

PERFoRM-Compliant Components Applied in the Siemens Use Case

Area	Solution	Supplier*
Middleware	Industrial Middleware	HSEL
Technology Adapters	Oracle Database Adapter	Loccioni
	MS-SQL Database Adapter	Paro
Data Analytics	Data Mining	MTC
	Min-Max-Data-Mining-Toolbox	TUBS
	Bayesian Diagnostics and Prognostics	Lboro
Visualization	Maintenance Task Editor and Schedule selection Tool	TUBS
Scheduling	Multiobjective Scheduling of production orders and maintenance tasks (bio inspired)	UniNova
Simulation	Simulation Environment	Siemens

*Partners of the project (EU HORIZON2020 FoF PERFoRM 2015-2018).

defined by the PERFoRM system will be manually transferred to the existing SAP Advanced Planning and Optimization (APO) Scheduling.

Figure 10.4 represents an overview of the logical interaction between legacy systems and the new PERFoRM applications and Adapters.

10.4 PERFoRM-Compliant Components Applied in the Use Case

10.4.1 Data Analytics Services

The main task for the definition of maintenance tasks is the acquisition and representation of the current conditions of a machine to the maintenance personnel, who can then investigate the situation and decide on actions to be taken. For providing the necessary functionality, two different approaches have been chosen in order to cover a broad range of information available to the users of the overall system: The first approach is based on the use of Data Mining and the Bayesian Diagnostics and Prognostics Services (see Chapter 7). It mainly applies big data analytics techniques to data acquired by the machines and the production and Machine Data Acquisition System (Machine Data Acquisition (German: Maschinendatenerfassung (MDE))/(Process Data Acquisition (German: Betriebsdatenerfassung (BDE)), respectively (see Chapter 7). The data available is alarm data originated in the machine controls and failure ticket data that was manually entered into the maintenance management system in case of machine breakdowns or severe failures. Although the data available for analytics is limited in terms of representing physical conditions of the machines, this approach allows one to base the analysis on all information available since the machines have been put to operation, many years ago.

As a second approach, the Min-Max Data Mining toolbox (see Chapter 7) was chosen in order to be able to conduct the condition analysis for production machinery on actual physical measurements of the machines. Expectation wise, this approach shall allow personnel to achieve much more detailed results in order to identify wear, and therefore typical predictive maintenance issues, but it comes with the limitation of having no historic data available. Consequently, since the measurement and analysis devices have been developed and applied only during the runtime of the use case implementation, the statistical relevance of the results will only have reached a high significance at a time beyond the scope of the project.

Independently from the approach used, the three services are used to represent information on the current machine state (in terms of machine health) to the user. This is done by having various visualizations integrated within the visualization service, which allow the user to browse through the data generated by the services.

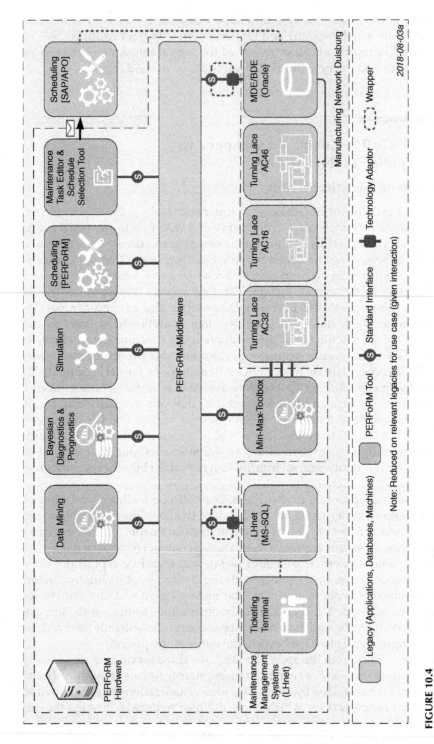

FIGURE 10.4

Architecture for the Siemens Use Case implementation.

10.4.2 Data Adapters and Middleware

In the chosen approach for the demonstrator realization, the integrated solution relies on accessing data from the legacy systems of the current factory. The first portion of required data concerns failure notifications and alarms produced by the tooling machines and employees using the maintenance management/ticketing system. This portion allows the maintenance personnel to investigate the current machine conditions in larger detail. The second portion is the planning data: The manufacturing tasks scheduled for the considered time frame in the future that have to be rescheduled once additional maintenance tasks have been defined.

Planning data is stored in the so-called MCIS (Motion Control Information System) Oracle Databases together with the machine failure and alarm data. The maintenance tickets are stored in the MS-SQL Database of the Maintenance Management tool. Naturally, both databases use data formats not identical to the PERFoRM information models, and exchange formats are consequently not compatible with the interfaces to the PERFoRM Middleware. With the two data adapters developed for the use case, these interfaces are provided REST endpoints to the data consuming services (see Chapter 5). When addressed, the data adapters generate a query to the legacy databases and convey the query result to the respective service, after translating the retrieved data into the fitting formats. Thus, the data adapters establish the connection of the PERFoRM System with the legacy system, providing data access to the legacy system for all services connected to the Middleware.

As one basic development of PERFoRM, the Middleware itself is used to provide the basic functionality necessary to run the PERFoRM-compliant System as a whole. Following its purpose, its major task is found in enabling the communication of the applied services.

10.4.3 Scheduling and Simulation

In contrast to the existing scheduling system, the handling of (planned) maintenance tasks differs in the PERFoRM approach. While in the existing system, a time slot to execute maintenance jobs is set in a fixed manner by limiting a machine's capacity for a certain time frame (which leads to limiting the amount of manufacturing tasks that can be assigned in that time frame to the resource), with the PERFoRM approach, a more flexible approach is chosen. By allowing the shifting of maintenance tasks in time, it is expected to improve the overall, combined performance of maintenance and manufacturing completion itself.

Therefore, maintenance tasks are defined in a similar manner like actual production tasks by a user. The core information that is required includes time restrictions as well as estimated efforts (earliest start, latest end, estimated duration) besides specifications on the factory item they shall be applied to (machine, potentially specified per subsystem).

However, reality almost always deviates from what was planned due to unforeseen events during manufacturing execution. Causes are smaller failures, quality issues and resulting rework, machinery problems, or simple uncertainties in the exact time planning and duration of the task to be executed. Even small deviations, especially in early production stages, might induce larger delays. At the same time, early deviations provide good potential to be fixed during the remaining time left until the production of the whole product is completed.

Deviations as described could in theory be considered during the scheduling process itself. However, for the given case, with a rather large number of tasks and resources to be scheduled, they are considered as too complex in order to generate adequate results in the available time for schedule generation.

Logistics Simulation, in contrast, can be used to overcome these drawbacks by analyzing the expected outcome of applying a certain schedule to given starting conditions. By running a large number of simulation experiments (with randomized statistical appearance of deviations as mentioned above), a statistically sound result can be achieved, indicating how robust a certain schedule is with respect to achieving a considerably good performance in total.

Therefore, the approach chosen is to generate several schedules, following common scheduling strategies, such as earliest or latest possible execution of the maintenance tasks (see Chapter 7). The schedules generated are then fed to the simulation, which is controlled by the simulation wrapper, which also constitutes the functionality to communicate with the other services through the Middleware. The simulation model covers the whole production area of the Duisburg plant. For all machinery, typical parameters like Mean Time Between Failures (MTBF), Mean Time To Repair (MTTR), and similar are statistically acquired or estimated in case no statistics are available. Changes in the availability and failure probabilities for the machines considered in particular have been estimated and are implemented in the current model.

10.4.4 User Interfacing Tool

The PERFoRM Demonstrator is designed to use an integrated user interface for all services, realized as a User Interfacing Service itself. It is realized as a website, thus in principle accessible from all computers within the factory network, and without the need to install and start an individual application per task. The visualization application incorporates views for the three main areas relevant in the use case: (i) The Data Analytics Services, (ii) the Task Definition, and (iii) the representation of the simulation results, and thus the schedule selection as depicted in Figure 10.5 (Data Analytics start page) (see Chapter 7 for more information about the visualization tool).

As described above, the Data Analytics part allows the user to navigate through the different data representations of the data analytics services' results. The user—expectedly a maintenance operator or manager at this

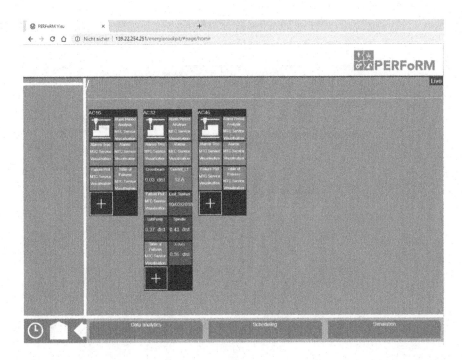

FIGURE 10.5
User Interfacing Tool—Landing Page view.

point—is provided with different data representations, such as diagrams of failure and alarm frequencies, alarm type distributions, and several others. It is possible to filter the results shown, for example, by start and end dates, machines, etc.

With the information available, the user is for the second step presented a task definition interface where he can manage maintenance tasks by defining times and durations, as well as subsystems and similar information for the scheduler itself.

Finally, the overall results are presented to the user as scheduling and simulation results in the form depicted in Figure 10.6.

10.5 Validation of the Use Case and Achieved Benefits

The Siemens use case demonstrator acted as a proof of concept for conversion of legacy equipment into Smart Components, which is a decisive stepstone toward the realization of the PERFoRM vision. By being able to integrate disparate databases into single accessible points, opportunities for loss of information could be reduced. The implementation activities lead to an increased

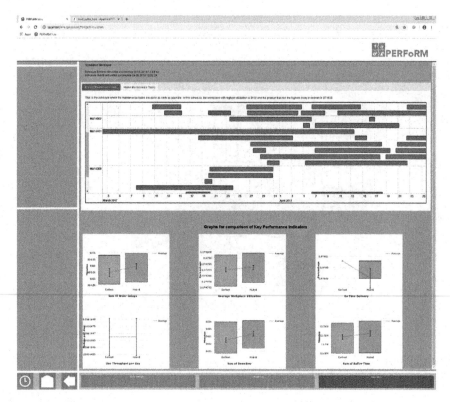

FIGURE 10.6
User Interfacing Tool—Simulation results and Schedule selection.

awareness and appreciation for Industry 4.0-compliant solutions and predictive analytics among various stakeholders of the factory, which tackles the challenge of industrial acceptance (see challenges in Chapter 1).

Because of the improved ability to diagnose faults prior to occurrence, a reduction of unplanned downtime as well as a reduction in root cause analyses and fault identification in unplanned downtime is expected.

Currently ongoing activities focus on the finalization of integration efforts (e.g., integration of all services and data sources, testing activities), trainings of respective roles (maintenance, planning manager), and the establishment of novel processes in the factory.

Also, a large-scale deployment of the concept by including additional machines of the shop floor provides further opportunities for the PERFoRM approach, since the PERFoRM Demonstrator covered only a small share of the shop floor machines.

Further potential lies in the integration with other internal digitalization projects in order to fully tackle the potential of Industry 4.0. An important prerequisite in this context is to improve the data situation and quality by adding sensors to machines.

10.6 Conclusion

The Use Case Compressors allow the application of the PERFoRM approach when it is applied to the Siemens Compressor Factory located in Duisburg, Germany, supporting the migration of these manufacturing premises into a digitalized, harmonized, flexible, and reconfigurable system according to the Digital Factory paradigm. The process of deploying and integrating PERFoRM-compliant components addressed in Chapters 4–7, as well as implementing both architectural and behavioral aspects associated to the PERFoRM architecture presented in Chapter 3, has been described in this chapter. The increased data quality and transparency improved the ability of predictive fault identification and contributed to a reduction of unplanned downtime and fault-finding effort.

References

Cala, A., M. Foehr, D. Rohrmus, N. Weinert, O. Meyer, M. Taisch, et al. 2016. "Towards industrial exploitation of innovative and harmonized production systems." Proc. of the 42nd Annual Conference of the IEEE Industrial Electronics Society (IECON 2016), pp. 5735–5740.

EU HORIZON 2020 FoF PERFoRM. 2016. "Deliverable D7.1: Siemens description and requirements of architectures for retrofitting production equipment." Unpublished manuscript, last modified October 10, 2018. http://www.horizon2020-perform.eu/index.php?action=documents.

"EU HORIZON2020 FoF PERFoRM 2015-2018." Accessed October 10, 2018. https://cordis.europa.eu/project/rcn/198360_de.html.

HORIZON2020 FoF Project. 2018. "PERFoRM." Accessed October 2018. http://www.horizon2020-perform.eu/.

Siemens "Industrial Compressor Manufacturing". 2018. "Business unit "Industrial Compressor Manufacturing"." Accessed October 10, 2018. https://www.energy.siemens.com/hq/en/compression-expansion/.

11

Use Case: White Goods

Pierluigi Petrali, Arnaldo Pagani
(Whirlpool EMEA)

Filippo Boschi, Giacomo Tavola, Paola Fantini
(Polimi)

José Barbosa, Nelson Rodrigues, Adriano Ferreira
(Instituto Politécnico de Bragança)

Giacomo Angione
(Loccioni)

CONTENTS

11.1 Introduction

Whirlpool (Whirlpool Corporation 2018) is a world leader in large domestic appliances with a strong presence in Europe with 20 factories and development centers located in six countries. Whirlpool Europe is a wholly-owned

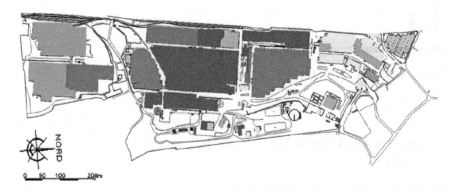

FIGURE 11.1
Biandronno site map.

subsidiary of Whirlpool Corporation, world leader in large domestic appliances. Its main sector is the White Goods one, and for this reason, it was chosen to focus on one of them.

Production harmonizEd Reconfiguration of Flexible Robots and Machinery (PERFoRM) demonstration activities were located in Biandronno, the Whirlpool largest site in Europe having a total area of 1,166,120 square meters, employing more than 2000 people, and encompassing three production units specialized for built-in appliances: refrigerators, traditional ovens, and microwave ovens, with a total yearly production of more than 2,300,000 pieces (see Figure 11.1).

The factory selected for the experimentation and demonstration is the Microwave Factory (see Figure 11.2). It is a pretty new production unit, built in 2014 from brown field, by transferring products and part of the production lines from the Swedish factory of Norrköping. The production transfer gave the chance to apply several innovations in the process both from organizational and technological point of view.

The factory is producing more than 450,000 pieces a year divided on four families: MIDI (see Figure 11.3), Mini, Opera, and Phoenix.

The production process is composed of eight major steps (see Figure 11.4):

1. Material incoming: Raw material (steel and stainless steel coils) and mechanical, electrical, and electronic component are delivered at any hour of the day every day. They are checked and stored in a warehouse.

2. Cavity line fabrication: The process aims to produce the internal cavity of the oven starting from carbon steel then painted or stainless steel in three major steps:

   a. Cavity parts stamping: Cavity parts (top and wrappers) are stamped in hydraulic presses.

   b. Cavity assembly: Cavity parts are welded to form an open box.

   c. Painting: Cavities made in carbon steel are washed and then painted.

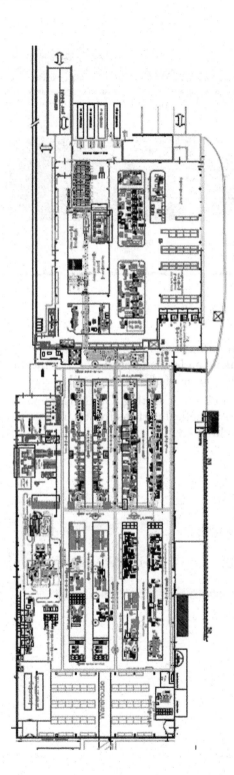

FIGURE 11.2
Detailed layout of Microwave Factory.

FIGURE 11.3
Built-in microwave of the MIDI family.

3. Door fabrication and assembly: Door is made of a combination of steel frame, a shield, a glass, and a handle. The most critical step in this process is the gluing of glass to the steel frame.

4. Part fabrication and silk screening: Other metal parts are stamped from steel coils. Aesthetical parts as front panel are marked using a silk screening technology.

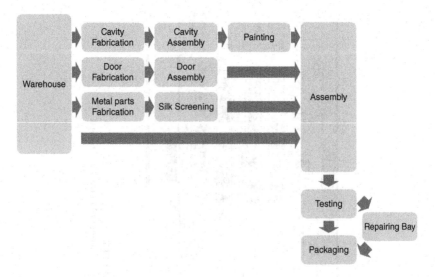

FIGURE 11.4
Materials and techniques used.

5. Final assembly: Cavity, door, and all other components are assembled in sequence in a continuous flow conveyor.

6. Testing: Ovens are tested from safety and functional point of view.

7. Repairing bay: Products not passing the test are examined and repaired in a separated bay.

8. Packaging: Ovens are finally packaged and sent to finished good warehouse.

From the logical point of view, the factory is organized according to the Whirlpool Production System (WPS): The material and information flows are described through the Value Stream Map (VSM), which highlights the relationship between production blocks in terms of complexity, buffers, sequence, timing, and resources (see Figure 11.5).

11.1.1 Continuous Improvement Process

White goods factories are always under the pressure of optimizing their performance and reducing their cost at the same time: Domestic appliances

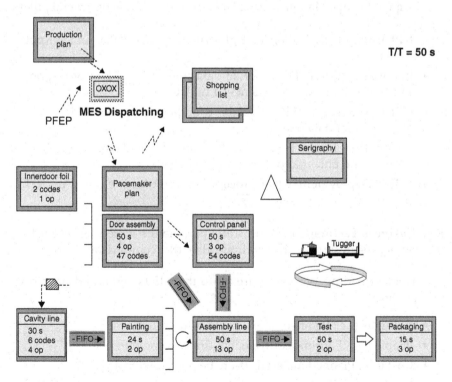

FIGURE 11.5
Value Stream Map of MINI family (incomplete).

are considered mostly commodities, and, in general, the business can provide only very low marginality. As a consequence, factories must put themselves in a continuous improvement process driven by cost reduction that turns out in a pipeline of activities aimed at reconfiguring the factory itself to better respond to external input as new product introduction or product reengineering, changing market request, technology change (e.g. Waste Heat Recovery (WHR) technologies), and so on.

Whirlpool Europe, being the world leader in White Goods appliances, is well representing this status, and the development of a Reconfiguration Support system will help and improve the current reconfiguration process.

11.1.2 KBF & KPI

The defined **Key Performance Indicators** (KPIs) structure focuses on operational KPIs, accompanied by existing financial KPIs. These operational KPIs on region, product group, and factory level allow tracking and steering of the specific WPS implementation and manufacturing operations in general.

Whirlpool Europe Manufacturing KPIs structure is built on several pillars:

- **KPI Cockpit**: Definition of KPI structure, core KPIs, and output format.
- **Reporting System**: Definition of data source (IT queries or responsible KPI owners) and frequency of updates.
- **KPI Tool**: Access and Excel database for data gathering and Internet-based tool for data output.
- **Calculation Logic**: Detailed definition of calculation logic, data sources, and potential pitfalls for each KPIs.
- **Policy Deployment**: Clear process for target setting based on KPI structure.

Key Business Factors (KBFs) are static parameters which represent how the factory is designed. They allow also the comparison of different factories.

KBFs identified by Whirlpool are more than 150 overall and have been grouped in:

- **Volumes** (e.g., Planned units to be produced in a year)
- **People** (e.g., Fulltime equivalent of direct workers)
- **Cost** (e.g., Planned material cost to be processed in a year)
- **Quality** (e.g., Cost of repair workers)

- **Inventory** (e.g., Days of stock)
- **Area** (e.g., Total factory covered area)
- **Maintenance** (e.g., Auxiliary material on stock)
- **Organization Setup** (e.g., IT costs)
- **Utilities** (e.g., Total HVAC costs)
- **Productivity** (e.g., Saturation of equipment)

Starting from the current status analysis, the main disadvantages of the processes currently in place, which should be improved, are listed below:

- The global view of the entire factory is not completely supported.
- The correlation between the static parameter KBFs and the dynamic parameter KPIs needs to be better identified and finalized.
- The simulation and the structured analysis of the whole factory behavior have to be improved.
- KPIs are evaluated off-line on a monthly/weekly/daily base, but not real time.
- Only some of the dynamic factors are monitored on real time.

The main objective of the Whirlpool Use Case, applying the PERFoRM approach, can be summarized as follows: To realize a real-time monitoring system able to correlate dynamic behavior of the factory to its KPIs implementation and to static indicators such as KBFs.

It is represented in the Figure 11.6.

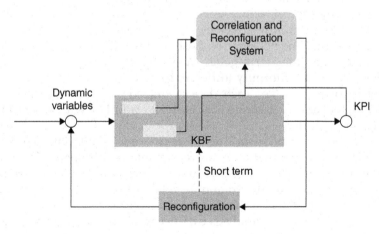

FIGURE 11.6
Closed loop schema of reconfiguration as a continuous improvement.

The model objectives were essentially consisting in a particular and dynamic system which is able to manage the following actions:

- To recognize the input: The variable external issues (e.g., market demand).
- To identify the KBF.
- To define a relevant model.
- To evaluate the output, that is, the KPI value.
- To compare inputs, KBFs and KPIs in order to choose the actions needed to obtain the system reconfiguration and to improve the output.
- To apply a sensitivity analysis, aiming at defining the relationships (i.e., sensitivity factors) among KBFs and KPIs.
- To validate the model consistency.

11.2 PERFoRM Architecture

For the present Use Case, the generic PERFoRM architecture (further details are given in Chapter 3) needs to be instantiated accordingly with the factory plant dedicated to produce microwave ovens. Therefore, it must be considered that the current factory continuous adaptation and medium term reconfiguration mechanism is based on a set of processes (Factory Master Plan, Profit Plan, Cost Deployment) which aim at improving KPIs through modification of factory assets and organization described by KBFs. Naturally, KPIs are mainly driven by shop-floor data of each single facilities and departments. The behavior of these facilities is monitored in order to meet middle- and long-term goals. Currently, the data gathered at shop-floor level lacks of uniformity (different formats and sources of data) and, moreover, correlation (i.e., each data is treated and analyzed without a model or a tool able to describe how each KPI is linked with others KPI or input factors or KBF).

The Use Case aims on installing a real-time shop-floor data acquisition to be able to react immediately on reasonable requirements in production planning. For this purpose, the data acquired and collected from the shop floor should be analyzed and correlated to extract, in advance, the KPIs and also to detect as earlier as possible disturbances or performance degradation. This analysis can be complemented with simulation facilities.

Furthermore, the PERFoRM generic architecture also needs to accommodate the introduction of innovation at the machinery level in the sense that an automatic leakage station, using a collaborative robot, needs to be integrated in such a way that the designated test is performed automatically and dynamically independently of the production orders currently being executed.

11.3 Architectural Mapping

Considering the particularities of the Use Case, the general system architecture is mapped as illustrated in the Figure 11.7, where generic blocks were updated by the legacy HW devices and SW applications covered by the Use Case.

As shown, the data acquired from the shop floor and currently collected by several databases are integrated in the PERFoRM ecosystem by the Middleware and in a PERFoRM database. The integration of these legacy databases requires the use of proper technological Adapters to transform the native data format into the data model defined by PERFoRM (see Chapter 4 for further details).

Several new tools are considered to provide the required advanced features, namely the monitoring and visualization system to support the online visualization of KPIs, the KPIs optimization tool that uses two different

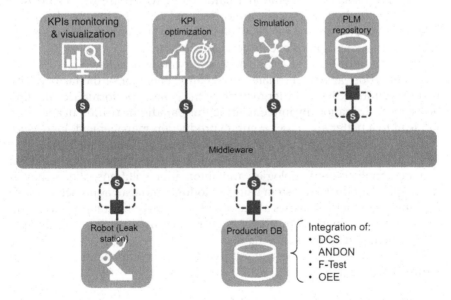

FIGURE 11.7
Architecture mapping for the Whirlpool Use Case.

models (MPFQ-K model and VSM) to identify strategies to improve KPIs and also considers a simulation tool having what-if game functionality to support the analysis of the impact of several degrees of freedom (DoF) in these KPIs.

The human-machine interaction is mainly reserved for key decision makers (e.g., production and industrial engineering managers), which will use the monitoring and visualization tool, as well as the proper designed user interfaces for the simulation tool, to understand the current system performance and to study how KPIs can be optimized.

11.4 Simulation Tool

The challenge of this Use Case is to solve the lack of available real-time shop-floor data and its integration into the production system controlling and planning.

Currently, the company monitors the behavior of the factory by measuring some KPIs that are then used to take decision on both dynamic and static factors. It monitors also the behavior of a selection of single machine or equipment measuring some of their performance in order to put in place local optimization actions. This approach gives only a snapshot of the situation on the shop floor at one specific moment, as it is based on a static (linear) model. This means the operational parameters, which are the results of a dynamic (evolutionary) process, are considered only as given by definition when instead they are the results of some factors chosen a priori and of the sequence of asynchronous events generated on the field.

The intent, indeed, is to establish a new paradigm, called Value Stream Reconfiguration, able to correlate the dynamic behavior of the factory with its static factors (KBFs) in order to forecast the general performance allowing its fast reconfiguration. By implementing this paradigm, a different approach will lead to a better and more accurate production management, increasing productivity, reconfigurability, and real-time control (Cala et al. 2016).

In fact, Value Stream Reconfiguration, using the combination of simulation activities, analysis methodologies, and automation solutions, allows modeling a typical production system using a modular and lean approach to make the simulation methodologies aligned with "fast reconfiguration" paradigm that could be influenced by the changeable nature of current production environment and by market demands.

The simulation tool can be split in three different components as shown in Figure 11.8:

1. **Simulation Automa**—needed to assess the *operational performances of individual machine.*

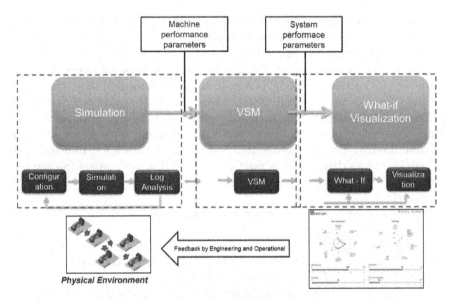

FIGURE 11.8
Value Stream Reconfiguration.

2. **Simulation VSM**—needed to monitor the *strategical performances of overall production systems.*

3. **What-If Analysis and Visualization**—needed to identify the *eventual changes to be applied, to provide the feedback for readjusting the physical plant, and, therefore, to adopt the "Reconfiguration" process.*

11.4.1 Simulation Automata

The simulation activity used in this Use Case aims at describing the process of single workstation involved within Whirlpool shop floor. For this reason, its objective is to provide, starting from machine technical parameters, more specific and operational details of each workstation once it is involved within the production context. Instead, their mutual relationship, needed to depict the overall production system behavior, will be evaluated by implementing Simulation VSM (described in the next paragraph).

In order to carry out this activity, two main concepts are taken into account.

The first consists in considering the domain of time as a set of discrete temporal instants. This means that the proposed approach refers to a Discrete Event Simulation (DES) as it considers a physical system representation based on a mathematical/logic model able to point out state changes at precise points in simulated time and when a specific event occurs (Nance 1993). It is one of the four main methods in simulation modeling described that is distinguished together with the AB by considering precisely the change of state as discrete temporal steps as well as by the level of abstraction a method can represent (Borshchev and Filippov 2004).

This ability is ensured as the model changes its state every time an event occurs and as the model is able to track the temporal sequence of the occurred events and associated state transition. In this way, the model can provide information about the state history. Hence, it is possible to estimate time-based performance indicators associated to each state. In fact, as described by Boschi, Tavola, and Taisch (2017), knowing that the time is considered as discrete, it is possible to describe some machine parameters as an event the machine is subject to, constant over time intervals and memoryless.

The second aspect refers to main production logics with which a productive system can be managed.

In fact, as stated in literature (Schonberger 1982), a discrete flow of material can be regulated by different production logics that can be distinguished in a pull or push logic control system. The characterizations of the push/pull distinction that have appeared in the literature (González-R, Framinan, and Pierreval 2012; Nahmias 2009) can be summarized into the following definition:

In a pull system production is triggered by actual demands for finished products, while in a push system production is initiated independently of demands and usually based on forecasting (Liberopoulos 2013; Tavola, Taisch, and Boschi 2017).

Following these aspects, it is possible to model the discrete production flow of Whirlpool Use Case as a composition of different entities (machines, buffer, etc.) that follow the Push/Pull logic in a discrete time domain. In this context, different workstations, denoted by WSi, $i = 1,..., n$ separated by buffers of finite capacity, denoted by Bi, $i = 1,..., n$ can be represented with different standard entities. Each entity is considered as a mathematical abstraction that describes all the states representing each possible situation in which these entities may ever be and all inputs and outputs with defined events. This assumption leads to describe a sequential logic as a number of output (n,o) which depends on the present and the past values of the input (n,i) and to formalize their behavior as a Finite State Machine (FSM).

In this way, it has been possible to propose a reduced set of independent standard entities able to represent each actor involved within a typical production line (i.e., machines, buffers, etc.) and to instantiate the system states changes in a discrete time steps using Finite State Automata (FSA) formalism (Tavola et al. 2017).

The set of standard entities has been created using the Simulink software (Simulink® User's Guide. MathWorks® 2016) and formalizing a special library from which it is possible to draw in order to compose and model any production plant. They, as described by Tavola et al. (2017), are depicted in the Figure 11.9.

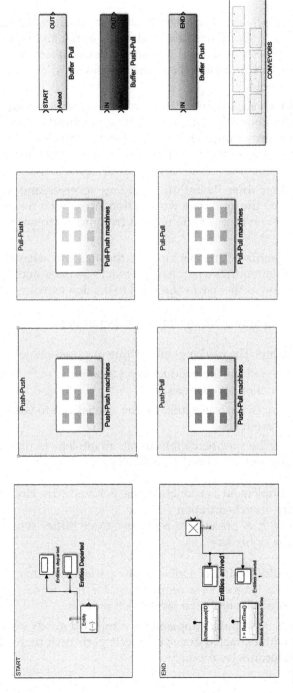

FIGURE 11.9
Simulink Library.

The Figure 11.9 (on the left) shows that a generic workstation consists of a machine that can be classified as:

1. **Push-Push Machine:** When the Push-Push Machine finishes the part, it pushes to the downstream station or buffer and starts working on a new part only when it is pushed from its upstream machine WS_{i-1}. If no part is available there, the machine is idle and the production is on hold.

2. **Pull-Pull Machine:** When Pull-Pull receives a request from downstream workstation WS_{i+1}, it pulls the request to upstream workstation WS_{i-1} for providing a component. Once it finishes the product it releases to the downstream requestor and waits for another request.

3. **Push-Pull Machine:** Push-Pull Machines receives and works a part pushed by the upstream workstation WS_{i-1}, but it releases that part only when it receives the request from downstream workstation WS_{i+1}.

4. **Pull-Push Machine:** This machine has the opposite behavior of the previous one: It requires a part as soon as it is available and it pushes the part as soon as that part is finished to the downstream entities.

All machine types operate with the following parameters:

- Operational time: The working time to transform/assemble products.
- Setup time: Based on the product type, a setup is executed every time the production lot changes.
- Worked products: Each machine can be specialized to work only a limited set of products.
- Optionally can be specified a failure rate to implement unavailability due to fails and associated repairing cycle.

All above entities implement a production phase characterized by an optional random variability of cycle duration.

In the same way, it is possible to represent three buffer typologies (see Figure 11.9—on the right), as:

1. **Pull Buffer:** Pull Buffer describes a raw parts buffer that provides material when it receives a request and for this reason is called "Pull." It is the first entity of a production process.

2. **Push Buffer:** Push Buffer implements a finished goods warehouse or a buffer with extracted items; it collects parts each time a Push-x workstation terminates production.

3. **Pull-Push Buffer:** These buffers can be used to model the Interoperational buffers. It is a passive intermediate entity receiving components and providing parts when they are requested.

All buffers instantiated are characterized by a given maximum capacity.

Following these definitions, it is possible to represent the discrete production flow of Whirlpool Use Case drag and dropping each of these standard entities from library, aggregating, and interconnecting them and specifying for each one the upstream station(s) or the downstream(s) entities. In doing that, it is necessary to respect compatibility rules that take into account how each entity works (i.e., input and output) (see Figure 11.10).

The simulation execution is carried out utilizing MATLAB platform as it provides SimEvents® (MathWorks 2016c. MathWorks®) as a tool that adds Discrete-Event Simulation (DES) capabilities to the Simulink® modeling platform (Simulink® User's Guide. MathWorks®). Therefore, providing a framework that helps easily express and compose DES systems in various graphical and textual programming paradigms, this platform is allowed to simulate a variety of application scenarios including modeling of real-time operating systems, operations research problems in manufacturing and logistics, and hardware modeling. Simulation activity provided different graphics and a specific Log needed to aggregate the punctual information of specific machine. In this way, it has been possible to figure out the operational parameters of each production assets involved within the production system (Figures 11.11 and 11.12).

Running Matlab code, it has been possible to figure out the performance parameters of production workstations. In particular, the analysis is focused on identifying the production losses of each machine that can be clustered, coherently with academic literature (Ron and Rooda 2006), in three different components:

- Loss of Availability
- Loss of Performance
- Loss of Quality

These components are needed to realize the Overall Equipment Effectiveness (OEE) that describes with unique value the operational performance of each machine and the other indirect and "hidden" costs, which according to Iannone (Iannone and Elena 2013) are those that contribute with a large proportion of the total cost of production (Bamber et al. 2003).

For this reason, this value can be considered as one of the factors impacting the actual production capacity and the Cassinetta possibility to achieve the predetermined production target. Therefore, it is needed to evaluate its

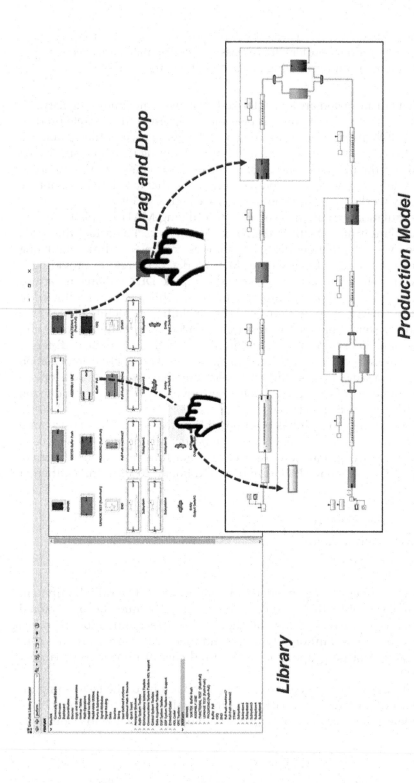

FIGURE 11.10
Value Stream model creation.

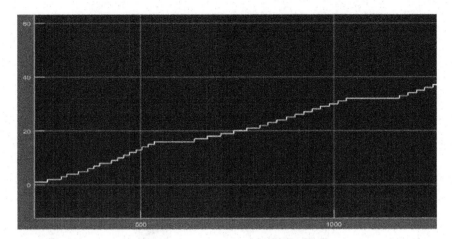

FIGURE 11.11
DES of a parameter.

effect on overall production system. This concept is described in the following paragraph (see Figure 11.13).

11.4.2 Simulation Value Stream Mapping

As described in the previous section, the main objective of VSM is to identify the congruence of production parameters and their impact in order to verify the production plant adequacy to achieve a determined production target. It, firstly, defines a particular production system model, combines the most significant KBFs with the relevant KPIs (considered in terms of production

Time [s]	Station	Event	Product ID	LOT	Scrap	Leak. tes	Func. Te
0	Assembly	Start_work	1	Lot 1			
30.46	Assembly	Start_work	2	Lot 1			
30.46	Assembly	Start_work	2	Lot 1			
60.92	Assembly	Start_work	3	Lot 1			
60.92	Assembly	Start_work	3	Lot 1			
78.79	Assembly	Start_work	4	Lot 1			
78.79	Assembly	Start_work	4	Lot 1			
94.9	Assembly	Start_work	5	Lot 1			
94.9	Assembly	Start_work	5	Lot 1			
111.01	Assembly	Start_work	6	Lot 1			
111.01	Assembly	Start_work	6	Lot 1			
137.08	Assembly	Start_work	7	Lot 1			
137.08	Assembly	Start_work	7	Lot 1			
157.02	Assembly	Start_work	8	Lot 1			
157.02	Assembly	Start_work	8	Lot 1			
172.82	Assembly	Start_work	9	Lot 1			

FIGURE 11.12
DES of production lot.

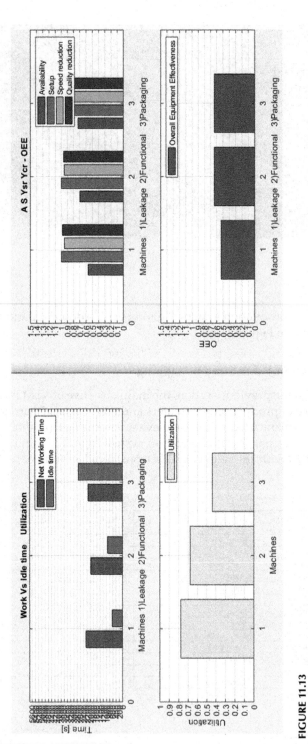

FIGURE 11.13
Overall Equipment Effectiveness (OEE) at a glance of whole production system.

aspect), and then allows the identification of the current status of shop-floor results. Doing this, it is possible to:

- Verify if the production plant is able to get the target.
- Describe how the plant can reach the production target considering the production rate (Throughput) and product lead time.
- Facilitate the what-if analysis based on the variation of several KBFs and support decision-making strategies.

The configuration of Value Stream is based on a visual method that uses predefined notations and symbols to assess the value and to determine the workstation process (Rother and Shook 2009). To this aim, it has been assumed that a table represents each workstation where KBFs (selected by process owner or whatever user that have responsibility to overall production plant, such as plant manager or head of department), simulation results, and KPIs are combined such as depicted in Figure 11.14.

The analysis of VSM can provide different outputs.

The first one is the evaluation of Overall System Performances needed to understand if the production target can be reached and how efficiently. For this reason, the following parameters are evaluated:

- Throughput
- Takt Time
- Theoretical Production Time
- Total Production Time
- Total Actual Lead Time
- Stock (pcs)
- Stock (hours)

The second aspect is the analysis of the actual machine ability to reach the target and the effective production balance and production flow as shown in the Figure 11.15.

As the Equipment Efficiency refers thus to ability to perform well at the lowest overall cost, the last output is the analysis of the conditions that could limit the production capacity considering

- cost analysis,
- loss of production,
- repair cost, and
- warehouse cost.

The third component of the simulation tool deals with the visualization of the overall results, highlighting the most strategic. In particular, by

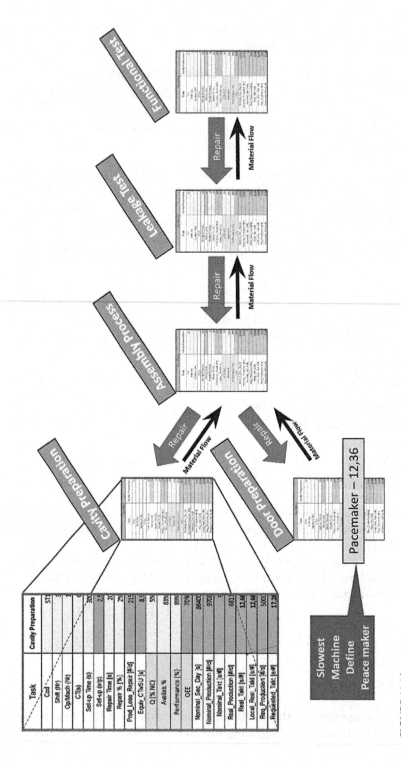

FIGURE 11.14
Value Stream Mapping.

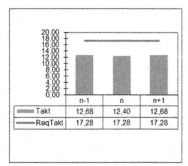 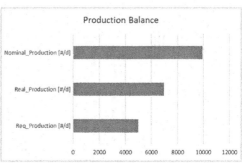

FIGURE 11.15
Production Balance.

containing significant product-specific information about all phases of the production process and by including order-specific information, such as the product price and production costs, the evaluation of each production stage can be generated and visualized with very little effort. Furthermore, using a web-based visualization tool, by KPI monitoring and by performing what-if analysis based on the variation of several KBFs, support decision-making strategies can be provided (see Figure 11.16).

11.4.3 KPI Monitoring with What-if Game Functionality

Used by many industrial sectors, the Six-Sigma is a set of techniques and tools for process improvement (Montgomery 2009), targeting the variability reduction in key product quality indicators, whose goal is the achievement of a very low level of defects. Assuming that the improvement of process performance and a reduced variability of the key parameters of the running manufacturing processes are critical features, the module is responsible for the statistical quality control of the temporal evolution of the KPIs, to ensure the expected quality of the products or product parts in production.

To perform the online statistical process control, a control chart for each KPI, relative to each of the stations, graphically displays the evolution of the KPI over time. The control chart plots the KPI's data points against some control lines. A centerline representing the average value of that KPI is displayed. Three other pairs of lines surround this centerline at One-sigma, Two-sigma, and Three-sigma (i.e., at once, twice, and three times the variance of the KPI). All these lines are used to assess some pattern denouncing out-of-control condition. Notably the Western Electric Rules (Electric 1956) are used:

Rule 1: One data point falls outside the Three-sigma control limits.

Rule 2: Two out of three consecutive data points fall beyond the Two-sigma warning limits.

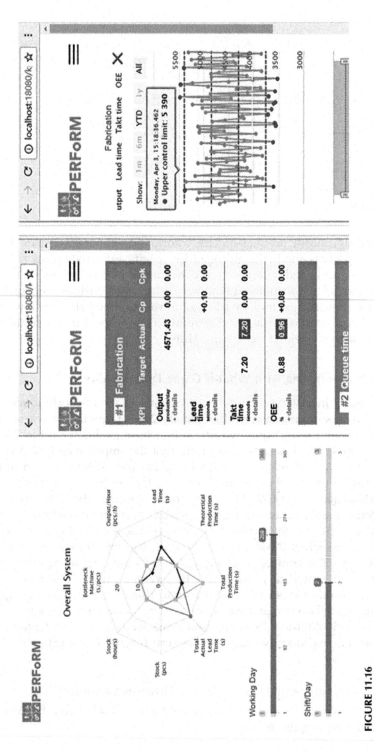

FIGURE 11.16
Key Performance Indicators (KPI) and Monitor tool.

Rule 3: Four out of five consecutive data points fall beyond the One-sigma limit, on the same side of the centerline.

Rule 4: Eight consecutive data points fall on one side of the centerline.

Additionally, the following rules were also considered (Montgomery 2009):

Rule 5: Six data points in a row steadily increasing or decreasing.

Rule 6: Fifteen data points in a row fall within the One-sigma limits (stratification).

Rule 7: Fourteen data points in a row alternating up and down.

Rule 8: Eight points in a row on both sides of the centerline with none falling within the One-sigma limits (mixture).

Points that present a pattern abnormality are signaled (e.g., be painted red) for better visualization. Additionally, the points that falls between the Two-sigma and the Three-sigma limits are stand out by the use of appropriate interval limit colors. In Figures 11.17 and 11.18, it is possible to see the result of the practical implementation of Rule 1.

11.4.3.1 Application at Whirlpool (WHP)

The what-if game mode, provided by this module, adds an extra functionality by introducing DoF intervals on the KBFs, allowing to foresee how the system behave within the specified intervals. Therefore, new production scenarios can be elaborated considering a more sustained decision-making process, based on the obtained expected KPIs. This mode is initiated by the decision maker for possible scenarios assessment or, for example, in order to mitigate the deviation of the actual KPI values from target values.

The simultaneous interval variation of several KBFs increases the scenarios combinatorial search space. Therefore, and without compromising the tool responsiveness, an intelligent scenario generation is implemented, prioritizing the calculation of the most promising scenarios (as shown in Figure 11.19), in a similar way to that proposed by Leitao, Rodrigues, and Barbosa (2015).

The What-if game Module uses internal info and the KPI Calculation & Statistical Quality Control Module to generate relevant outputs that must be provided to the Visualization Module, allowing the presentation of data in a way that could be evaluated at a glance.

The what-if game mode functionality presents data in a spider diagram, aggregating all the relevant KPIs into one graphical display (see Figure 11.20). The left side of the figure shows the overall system what-if KPIs results while, on the right side of the figure, the user can study other levels of granularity by searching KPIs at processing/station level.

The decision maker can change the desired KBFs by adjusting the sliders located at the lower part. With this, a new set of solutions will be displayed to the user allowing a thorough impact assessment of these adaptations.

FIGURE 11.17
Six Sigma CP CPK.

11.5 Reconfigurable Cobot

Whirlpool is really interested and motivated to minimize changeover effort and allow fast integration of automation systems into the wider company Information and Communications Technology (ICT) infrastructure. The reconfigurable robot application goes exactly in this direction.

In particular, this application considers the electromagnetic compliance (EMC) leakage test which is performed by a Cobot (Universal Robot UR10)

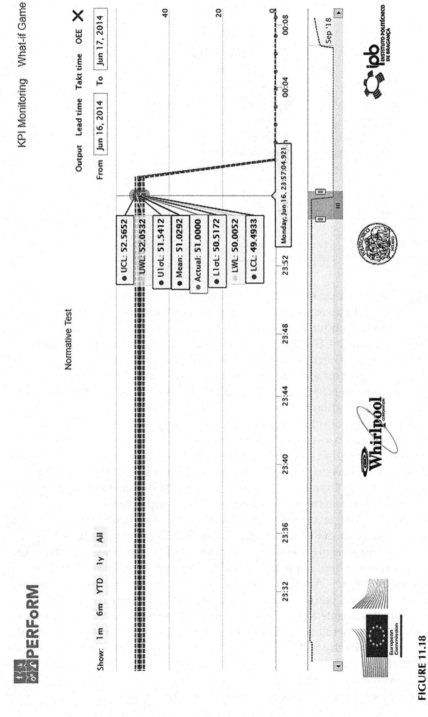

FIGURE 11.18
Bandwidth and time trend.

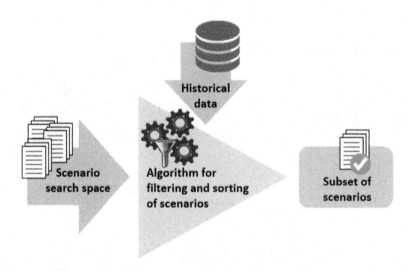

FIGURE 11.19
Scenario generation process.

equipped with a microwave probe which enables to detect microwave leakages. The Cobot moves the probe following a predefined path (Figure 11.21) with a speed of about 25 cm/sec. In order to correctly measure possible microwave leakages, the probe must be as close as possible to the surface of the oven and the direction must be perpendicular to it.

If the probe detects a microwave leakage beyond the limit of 0.8 mW/cm², the oven is sent to reparation. Figure 11.22 shows the Cobot installation and reachability simulation.

When a new model of oven is produced or when the detection path must be updated for any reason such as reducing cycle time, Whirlpool needs to stop the line and reprogram the Cobot. New Cobots, and, in particular, Universal Robots offer nice user-friendly programming features like the manual guidance and intuitive Graphical User Interfaces (GUI), but the time for programming and testing the robot movements is never negligible.

The Robot Adapter developed following the PERFoRM approach and described more in detail in Chapter 5 (Section 5.5.2) permits to reduce the time for programming the Cobot, but, most of all, to avoid stopping the production line when a new path must be defined or an existing path must be modified. In fact, Whirlpool product designers can draw the microwave leakage detection path using a Computer-aided Design (CAD) tool. The Robot Adapter is able to read the 3-D model exported by the CAD tool and automatically generates the proper robot program. This robot program can, then, be sent directly to the Cobot for being executed (e.g., in case of small adjustments), or first used in a simulation environment for validating the robot movements in order to avoid, for example, unreachable positions, robot singularities, or, even, collisions with the surrounding objects.

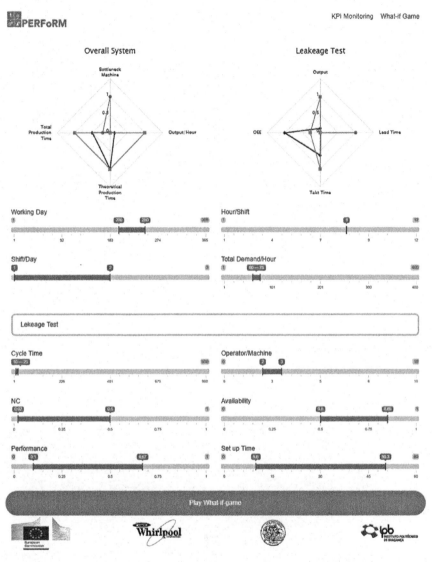

FIGURE 11.20
View of the tool in What-if Game mode.

The Robot Adapter has been fully integrated in the PERFoRM architecture and extensively tested and validated in the MTC laboratory verifying the correctness of the generated robot program and the robot path execution (Leitao et al. 2015).

The off-line programming method offered by the PERFoRM Robot Adapter facilitates the Cobot reconfiguration and permits to reduce the effort required when a new product needs to be produced. Considering the rapid change of

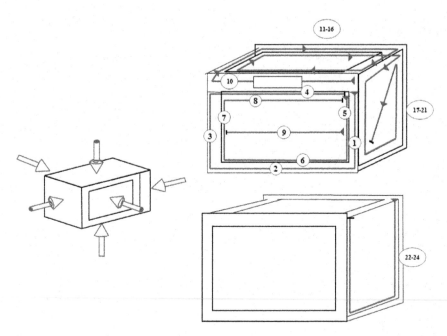

FIGURE 11.21
Leakage detection probe path and direction.

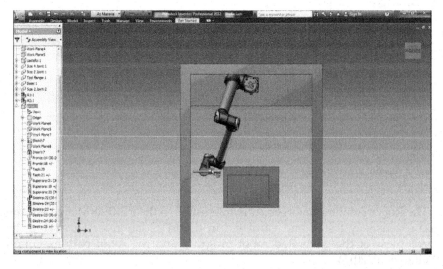

FIGURE 11.22
Cobot installation and reachability simulation.

the today's market demand and the consequent continuous development of new products, this turns out in a big advantage for Whirlpool. Moreover, the Robot Adapter permits to link two important manufacturing functions: Product design and product quality test. In fact, the product designer can draw the path for searching for microwave leakages by herself/himself, and since she/he designed the oven, she/he is the best person to know where leakages could occur.

11.6 Human Impact

The previous sections of this chapter have introduced the business needs and illustrated the technological solutions provided by the application of the PERFoRM approach. This section shed some light on the human involvement in the Use Case on two different levels.

The identification of the needs and requirements; the analysis, evaluation, and selection of the technological solutions; and the decisions leading to the adoption have been carried out by the Operations Excellence and Geographic Information Systems (GIS)-Manufacturing managers and engineers, who are also in charge of planning and managing the process of introducing the new technology in the manufacturing environment.

Besides technical implementation and deployment, the use and maintenance of novel technologies has to be integrated in the design or redesign of human activities. The users have to be identified, introduced, and, if necessary, trained to operate the tools; the maintenance strategies have to be defined.

In this Use Case, there are two main groups of users: the Value Stream personnel and the industrial engineers.

The Value Stream leader and operators use the automata and Value Stream tools to strengthen their systemic understanding of the manufacturing machines and systems. They increase their knowledge about the effects of individual variables, such as setup time, percentage of nonconforming products, etc., on the overall production performances. Therefore, they receive better support for the identification and prioritization of continuous improvement interventions.

Furthermore, the value stream personnel use the reconfigurable Cobot as an additional resource on the shop floor. They integrate switching-on/off, supervision, and a set of trouble shooting/emergency tasks into their daily operations.

The industrial engineers, and the value stream personnel, are the users of the what-if tool. They can use the tool independently or together to develop, analyze, and evaluate reconfiguration and refurbishing solutions for the manufacturing system.

The maintenance strategy includes training the value stream personnel of the "help chain" and GIS-MFG to enable first aid and simple trouble shooting together with service contracts with the technology providers.

Overall the introduction of the PERFoRM technology does not have a major impact on human resources and just requires few days of initial training for the above-mentioned groups. However, it can be seen as a step toward a more digitally empowered organization and personnel, as envisioned by the fourth Industrial Revolution paradigm, in which humans have a central role, increase their knowledge, and make informed decisions.

11.7 Business Impact

- Simulation has been used for decades as a tool to support decision-making in manufacturing systems. It is far cheaper and faster to build a virtual system and experiment with different scenarios and decisions before actually implementing the system. Simulation has been widely used to support decisions in manufacturing systems' operations and configuration.
- Nevertheless, knowing the changeable nature of manufacturing systems that is affecting the current production environment due to smart technologies adoption and market demands, the new approach to model a typical production system based on a modular and lean paradigm it's very helpful to make the simulation methodologies aligned with "fast reconfiguration" paradigm.
- The relationships among the KBFs, the drivers of the production system, and the relevant KPIs are crucial to better tune the manufacturing environment.
- The possibility to use Robot or Cobot automatically self-adapting and reconfiguring decoding the original 3D CAD file is opening new possibilities for automation.

11.8 Conclusion

As described, Whirlpool Use Case is split into two areas, apparently not linked. The first is addressing the overall setup of a manufacturing facility and the second is the automation level.

This was done by purpose in order to prove the approach in two different domain of ISA-95 enterprise architecture (ISA Organization 2018):

- Level 3—Reconfiguration of a Value stream based on KPI/KBF and simulation tool.
- Level 1—Reconfiguration at shop floor using Cobot.

The results achieved are very promising in both areas enabling to move to exploitation phase with a more robust industrialization and integration into IT architecture of the company.

References

Bamber, C. J., P. Castka, J. M. Sharp, and Y. Motara. 2003. "Cross-functional team working for overall equipment effectiveness (OEE)." *J. Qual. Mainten. Eng.* 9 (3): 223–238. doi:10.1108/13552510310493684.

Borshchev, A., and A. Filippov. 2004. "From System Dynamics and Discrete Event to Practical Agent Based Modeling: Reasons, Techniques, Tools."

Boschi, Filippo, Giacomo Tavola, and Marco Taisch. 2017. "A Description and Analysis Method for Reconfigurable Production Systems Based on Finite State Automaton." In *Service Orientation in Holonic and Multi-Agent Manufacturing: Proceedings of SOHOMA 2016.* Vol. 694, edited by Theodor Borangiu, Damien Trentesaux, André Thomas, Paulo Leitão, and José B. Oliveira, pp. 349–358. Studies in Computational Intelligence 694. Cham, S.I. Springer International Publishing.

Cala, A., M. Foehr, D. Rohrmus, N. Weinert, O. Meyer, M. Taisch, et al. 2016. "Towards Industrial Exploitation of Innovative and Harmonized Production Systems."

Electric, W. 1956. *Statistical Quality Control Handbook.*

González-R, Pedro L., José M. Framinan, and Henry Pierreval. 2012. "Token-based pull production control systems: An introductory overview." *J. Intell. Manuf.* 23 (1): 5–22. doi:10.1007/s10845-011-0534-4.

Iannone, Raffaele, and Maria Elena. 2013. "Managing OEE to Optimize Factory Performance." In *Using Overall Equipment Effectiveness for Manufacturing System Design,* edited by Vittorio Cesarotti, Alessio Giuiusa, and Vito Introna: INTECH Open Access Publisher. DOI: 10.5772/55322. London, UK.

ISA Organization. 2018. "ISA95." Accessed October 15, 2018. https://www.isa.org/isa95/.

Leitão, Paulo, Nelson Rodrigues, and José Barbosa. 2015. "What-If Game Simulation in Agent-Based Strategic Production Planners." In *2015 IEEE 20th Conference on Emerging Technologies & Factory Automation (ETFA): 8 - 11 Sept. 2015, City of Luxembourg, Luxembourg,* 1–8. Piscataway, NJ: IEEE.

Liberopoulos, G. 2013. "Production release control and the push/pull and make-to-order/make-to-stock distinctions." *SMMSO 2013.* 113–120.

MathWorks. 2016c. MathWorks®. " MathWorks. 2016c. SimEvents®, User's Guide. MathWorks®, Release R2016a, Natick, MA."

Montgomery, D. C. 2009. *Introduction to Statistical Quality Control.*

Nahmias, Steven. 2009. Production and Operations Analysis. *The McGraw-Hill/ Irwin Series Operations and Decision Sciences*, 6th ed., internat. ed. Boston, MA: McGraw-Hill.

Nance, R. 1993. "A history of discrete event simulation programming languages." ACM SIGPLAN Notices, 28 (3), pp. 149-175. Cambridge, Massachusetts, USA.

Ron, A. J. de, and J. E. Rooda. 2006. "OEE and equipment effectiveness: An evaluation." *Int. J. Prod. Res.* 44 (23): 4987–5003. doi:10.1080/00207540600573402.

Rother, Mike, and John Shook. 2009. *Learning to See: Value-Stream Mapping to Create Value and Eliminate Muda.* Version 1.4. A lean tool kit method and workbook. Cambridge, MA. Lean Enterprise Inst.

Schonberger, Richard J. 1982. *Japanese Manufacturing Techniques: Nine Hidden Lessons in Simplicity.* New York: Free Press. http://www.loc.gov/catdir/bios/ simon051/82048495.html.

Simulink® User's Guide. MathWorks®. "MathWorks. 2016. Simulink® User's Guide. MathWorks®, Release R2016a, Natick, MA." Accessed on October 10, 2018. https://it.mathworks.com/help/simulink/

Tavola, Giacomo, Marco Taisch, and Filippo Boschi. 2017. "A Standard Approach to Production Systems Modelling Based on Finite State Automata." In *2017 IEEE 15th International Conference on Industrial Informatics (INDIN): University of Applied Science Emden/Leer, Emden, Germany, 24-26 July 2017: Proceedings,* 1117–1122. Piscataway, NJ: IEEE.

Whirlpool Corporation. 2018. "Whirlpool Europe Srl." Accessed October 15, 2018. http://www.whirlpoolcorp.com/.

12

IFEVS Use Case

Riccardo Introzzi, Pietro Perlo, Marco Grosso, Sergio Pozzato, Marco Biasiotto, Davide Penserini, Gioele Sabato, Sandro De Pasquale
(Interactive Fully Electrical Vehicles SRL)

Massimo Ippolito, Pietro Cultrona
(COMAU SPA)

Manfred Hucke
(XETICS GMBH)

André Rocha, Ricardo Silva Peres
(UNINOVA)

Michael Gepp
(SIEMENS AG)

CONTENTS

12.1 Introduction

12.1.1 Microfactory: A Solution of Excellence

Nowadays, manufacturing is often done in high-variation/low-volume conditions; market advantages are clear: Better tailoring to specific customer demand, improved responsiveness, and lower inventory requirements. However, this scheme has traditionally returned lower quality output. The Microfactory, proposed in this use case for applying the Production harmonizEd Reconfiguration of Flexible Robots and Machinery (PERFoRM) approach and described in this chapter, addresses the low-cost flexible and agile production of a variety of safe, ergonomics, and high-performing low-footprint pure electric vehicles.

12.1.2 Business Opportunity

Increasing the number of cars or size is not a global sustainable solution; moreover, the footprint of a car is also of concern given the tendency to move from car ownership to car usership. By 2030, most of the world's population will be concentrated in cities. Assuming this trend continues, by 2050 more than 80% of the world's population will live in an urban environment. Cities are places of innovation; they are the drivers of our economy and places where wealth and jobs are created (ACEA 2018).

The above vision is supported by the observation that most cities are implementing new plans of urban mobility, coordinating public and private transport providers, logistics/freight companies, and emergency and traffic information services. Therefore, light electric vehicles and buses are the fastest-growing segments of urban mobility industry and semi- and fully autonomous vehicles are expected to expand first in the urban context (RAND Corporation 2018).

The electric mobility altogether leads to a mind shift in habits and energy use. While most of the attention is on electric conventionally sized vehicles (cars), the developments are at most addressing e-bikes, low-speed e-vehicles, and e-buses.

The highest market growth is registered for Low-Speed Four-Wheel Electric Vehicles (China) and kei e-cars (Japan) with an expected 80% to 100% Compound Annual Growth Rate (CAGR) in the next 5 years. Low-Speed EVs (LSEVs) currently represent an energy profile, such as public transport and two-wheelers. They are fast enough for getting around big and heavily congested cities. In China, low-speed electric vehicles are driving clean high-speed urbanization. LSEV sales experienced a considerable growth with more than 600,000 units sold in 2015 and 1.2–1.5 million in 2016. However, none among these LSEVs would meet European safety standards on crash tests.

In 2016 Mobility-as-a-Service (MaaS) companies drove 500,000 passengers per day in New York City alone (Dailynews 2018). That was triple the number of passengers driven in 2015. MaaS dramatically lowers costs compared with car ownership besides requiring no investment.

Cost-saving, energy-saving, and lower emissions are key factors in driving consumers to adopt MaaS.

All technology evolutions are converging toward a dramatic increase of various forms of MaaS platforms. In this competitive context, providers are springing up like mushrooms all over the world.

Large automotive manufacturers are entering MaaS, taking control of the largest platforms. They have made serious investments in mobility and autonomous technology companies in the past 2 years.

- GM's $500 million investment in Lyft (Bloomber-1 2016) and owns Cruise Automation.
- BMW's ride-sharing service, ReachNow (TESLARATI 2016).
- VW's $300 million investment in Gett (Bloomber-2 2016).
- Tesla is going to have its own ride-sharing platform (Business Insider UK 2016).

- Toyota Motor Corp. has backed Uber for an undisclosed amount.
- Daimler owns Hailo, MyTaxi, Taxibeat, and Ridescout.
- Ford acquired on-demand shuttle service Chariot in 2016, and then bought a majority stake in self-driving start-up Argo.AI for $1 billion.
- Large original equipment manufacturers (OEMs) want that MaaS will continue by mid-sized or large-sized cars whether conventional ICEs or EVs, but most of the demand is for urban mobility per which the current mid-sized or large-sized vehicles do not meet the real needs.
- All MaaS have in common the need of novel forms of vehicles and new approaches to produce them.

For the great majority of MaaS providers (not owned by large OEMs) the challenge is having their own fleet but none have the experience to make their own vehicles because

- making safe autos is a very challenging business,
- large investments are necessary to produce safe cars, and
- two years on average are needed to start car production.

LSEVs currently represent an energy-efficient alternative to conventional Internal Combustion Engine (ICE) cars. Some key reasons behind this remarkable development are their small size and low price, requiring low operation and maintenance expenses. In the Shandong region only, a hundred companies are addressing the introduction of new low-speed vehicles targeting high production numbers, but

- none among the known LSEVs models meet the minimal requirements on safety against frontal and lateral crashes and
- none among the known production plants is organized to meet automotive quality standards.

The low-speed sector is exponentially growing, but the fast expansion also comes with reliability and safety risks due to the scattered, small-scale nature of the LSEV manufacturing industry. There is a large opportunity to start the commercialization of safe and high-performing small-footprint EVs.

The problem to reduce the investments has to be tackled from a global system design point of view; the cost of the electric powertrain is decreasing very fast in the motor-powered electronic and transmission systems, as well as the maintenance cost, but, most important, the electric powertrain allows a mind shift in the design of the chassis where most of the investments to produce a new car are required. While most of the public debate is on bat-teries and electric powertrain, the major concern of the OEMs is still on the necessary investment to produce a new vehicle; this is the reason why even large OEMs share platforms to shorten the return of their investments; examples are Fiat and Ford for the 500 and the Ka (Automotive News—Europe 2015), Daimler, and Renault for the Smart for four and the Twingo (Automotive Manufacturing Solutions AMS 2015). To be cost-effective, an elec-tric vehicle has to be born electric (CARS, everything electric 2010; IDTechEx 2011) and from that point of view, the Microfactory concept is particularly important. It allows a radical reduction of the development time and of the necessary investments to make a new vehicle. The innovation proposed addresses the fast-growing market of electric urban mobility of people and goods.

Within the EU R&D platform on urban EVs, Interactive Fully Electrical Vehicles SRL (IFEVS) has developed a new approach representing a radical advancement with respect to the worldwide state of the art. The design of a chassis, based on a mix of advanced high-strength tubular steels, has been dem-onstrated to meet all most stringent EuroNCAP crash tests. The Microfactory addresses the next step proposing a low-cost and low-energy-intensity manu-facturing process, allowing a flexible response to the market demand of differ-ent vehicle architectures (passenger vehicles, pickups, temperature-controlled food delivery vehicles, taxis, etc.).

12.1.3 Specific Objectives and Expected Outcomes

The proposed Microfactory implements an efficient and cost-effective manu-facturing chassis platform that could be adapted to several typologies of low-volume specialized urban electric vehicles characterized by

- variable-demand manufacturing,
- high-mix manufacturing,
- manufacturing where non-recurrent engineering costs become a large portion of the overall product cost,
- rate-dependent production.
- flexibility (multiple product designs can use the same tooling and equipment),
- agility (process flow can switch among different product designs) and
- high organizational level (every final product could be associated to each components systems it integrates as well as to all produc-tion steps).

The developed methodology is (globally) parametric and applicable to most vehicle categories. It is based on breakthrough procedures that allow the design and development of novel safe chassis in less than one-tenth of the usual time and at a small fraction of the usual cost. Specifically, a completely new chassis can be designed and manufactured with very simplified templates that do not require complex automated systems before the operation of welding could start. Essentially the elements composing the chassis are designed and fabricated in such a way that they could be easily assembled (error-free) to form the chassis.

This leads to a radical reduction of the necessary investments: A variety of different vehicle architectures can be derived from a single chassis.

12.1.4 Seamless Vehicle Design

The chassis is designed and constructed applying novel procedures to produce extremely lightweight chassis in which a mix of super-high strength steel tubes are laser-cut in such a way that their assembly-alignment would be error-free. The assembly of the chassis is made by very simplified low-cost templates, no molds or stamps are necessary to produce the full chassis, the frame of the powertrain, and the suspension system. The methodology is scalable and adaptable to vehicles of different classes and uses. Quite different vehicle architectures could be developed from a single basic chassis whose state-of-the-art performance has been demonstrated both by simulations and physical crash tests. Acoustical and thermal insulation are assured by a novel and radical approach based on low-cost lightweight composite closing the chassis.

Chassis safety against crash tests is assured starting from existing designs which also consider the behavior of the battery pack against both lateral and frontal impacts.

12.2 IFEVS Use Case

Within the context described in the previous section and the framework associated with the PERFoRM approach, IFEVS aims at making available low cost, flexible assembly lines to rapidly start the manufacturing of safe, ergonomic, clean, and efficient vehicles customized to local needs (IFEVS 2018).

At present, an experimental manual assembly line has been deployed, with no legacy systems, to adapt to the new architecture that will be developed by following the PERFoRM approach. The assembly line is conceived for 50 vehicles a day over two shifts in agreement with the analysis carried out by COMAU (COMAU Spa 2018).

The entire Microfactory is designed with an area of 10,000 m^2. It is intended to be energy independent, since it can be powered by solar panels, as shown in Figure 12.1.

FIGURE 12.1
Microfactory design. (From IFEVS 2018.)

12.3 Vision

High-quality EVs will be produced in low budget and flexible assembly lines. IFEVS' objective is the creation of a Cyber-Physical Factory Environment, based on a decentralized cloud architecture that connects the physical systems to all the involved entities.

The envisioned solution increases the automation level and enhances quality and yield: It improves the efficiency and reproducibility of production. The new approach reduces rework of sub-modules and part rejections, leading to energy and raw material saving.

Lots can vary from few units, in the case of vehicles transporting temperature-controlled goods (e.g., medical items), to a tenth for special freight delivery vans and up to several hundreds for passenger vehicles. High flexibility is needed to process all vehicle variants in the same work islands; so far, the only convenient possibility is offered by human operators.

Few changeovers are necessary to set up templates and parameters on welding islands to pass from one variant to another; operations are automatized and drastically reduce the switch time. The low upfront investments necessary for the typology of vehicles addressed in PERFoRM (several variants and small lots) reduce the produced number of vehicles necessary to assure the return of the investments.

To mitigate the variability related to low-volume specialized productions, plug-and-play hardware and software have been chosen to implement a cloud-based Cyber-Physical System (CPS) architecture with services aimed at predictive maintenance methods. At full capacity, the factory will have a CPS architecture that self-adapts and self-optimizes, thanks to the feedbacks given by operators, quality engineers, and sensor subnetworks distributed on the working areas.

12.4 Aims

The addressed vehicle typology is intended to satisfy the effective needs of a great majority of people without compromises on

- safety
- automotive quality standards (reliability, traceability),
- ergonomics,
- smartness,
- aesthetics, and
- costs.

The cost of industrial molds and tools to manufacture components is essentially the same whether the production is 2–50 items a day; a larger production needs several molds replica. Less than five vehicles a day would make it challenging to reach the break-even point and full return investments.

Manufacturing different vehicle architectures within the same day should only minimally influence the potential daily throughput even when small lots are produced. A large variability of vehicle mock-up demand is expected, particularly in the first couple of years of production.

All assembly steps follow the highest quality standards of the automotive industry; all subcomponents and assembly processes are tagged and logged to allow item and production-time tracking.

12.4.1 The Chassis

To fulfill the above-mentioned requirements, lots of efforts have been made to find a proper technological solution. A different approach than standard automotive turned out to offer several advantages: The design of vehicles is based on a tubular chassis made of seven sub-modules: Top/Roof, Front, Bottom, Rear, Sides, Doors, and Powertrain.

Figure 12.2 shows different types of vehicles divided into two subsets, each sharing a common chassis.

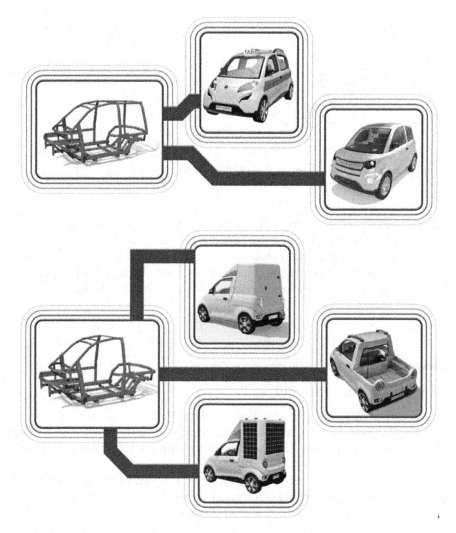

FIGURE 12.2
Variable design, common chassis of different vehicle models (IFEVS 2018).

Thanks to this modular approach and to the fact that the various types of vehicles slightly differ in the chassis size and components, it is easier to reach the desired flexibility of the assembly line.

The difference between the two above-mentioned chassis is evidenced in Figure 12.3.

Despite its low weight, a mix of Super High Strength Steel (SHSS) tubulars allows to meet all most stringent EuroNCAP crash tests on the chassis (EuroNCAP 2018).

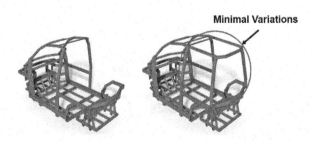

FIGURE 12.3
Comparison of pickup and passenger chassis (IFEVS 2018).

Each working island features its own template designed to hold only the specific items for a proper assembly and the following welding operations. In case of mock-up variants an automated sensor and feedback network monitors the production and provides blocking warnings in case of wrong operations. Components and sensors, identified by Xetics (XETICS GmbH 2018), are shown in Figure 12.4.

12.5 Microfactory Layout

The factory area is organized as shown in Figure 12.5.

In the Production Area, attention has been paid to the outline to reduce all the "Non-Added-Value" operations and minimize times related to item movements.

The central part of the Production Area is occupied by the Storage Area. This provides a common place for good delivery and storage as well as a quick supply to all the assembly areas with a minimal travel time.

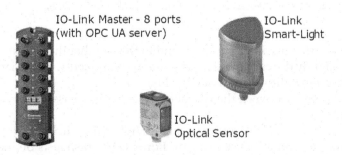

FIGURE 12.4
Core components of the automation network on the shop floor. (From IFEVS 2018.)

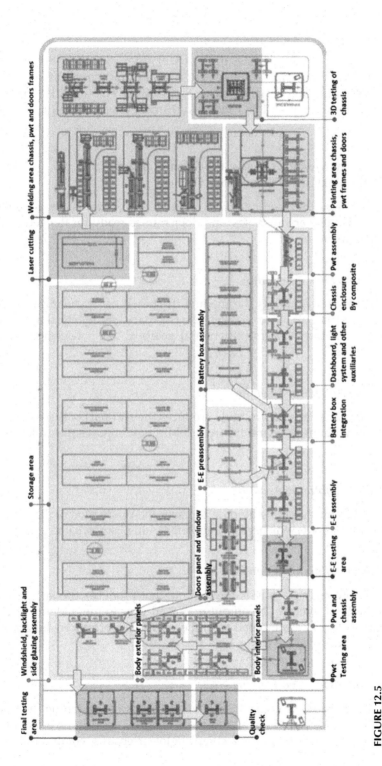

FIGURE 12.5
The assembly, from laser-cutting to final quality checks. (From IFEVS 2018.)

12.5.1 Laser-Cutting Zone

The laser-cutting module is the first part of the production process. In this area the SHSS tubulars are laser-cut with a Computerized Numerical Control (CNC) automated machine (see Figure 12.6). The result is a perfect fit among the parts that will be welded in the next phase of the production process.

12.5.2 Welding Area for Chassis, Precision Welding Technology (PWT), and Door Frames

This area receives the laser-cut SHSS tubulars and outputs the complete chassis, the two powertrain axles, and door frames ready for the geometry checks before painting.

12.5.2.1 Frame PWT Assembly

The frame PWT assembly is the module in charge of producing the two frames for the front and rear axles. These two parts are identical and they will host the motors, drivers, differential gearbox, and wheel hubs with shock absorbers.

12.5.2.2 Rear Subframe Assembly

The rear-subframe assembly module produces the rear part of the subframe that together with the floor and the front subframe composes the underbody of the car.

12.5.2.3 Floor Subframe Assembly

The floor subframe assembly module oversees the production of the central part of the underbody, which, once completed, will host the battery pack.

12.5.2.4 Front Subframe Assembly

The front subframe assembly is the module that produces the last part of the underbody of the car. The output produced goes to the underbody geo framing module, to be combined with the previously produced subgroups of the underbody.

12.5.2.5 Underbody Geo Framing

The underbody geo framing is the module in charge of combining the front subassembly, the central floor, and the rear subgroup and to give them the correct geometry. The output is a complete lower chassis, common for the

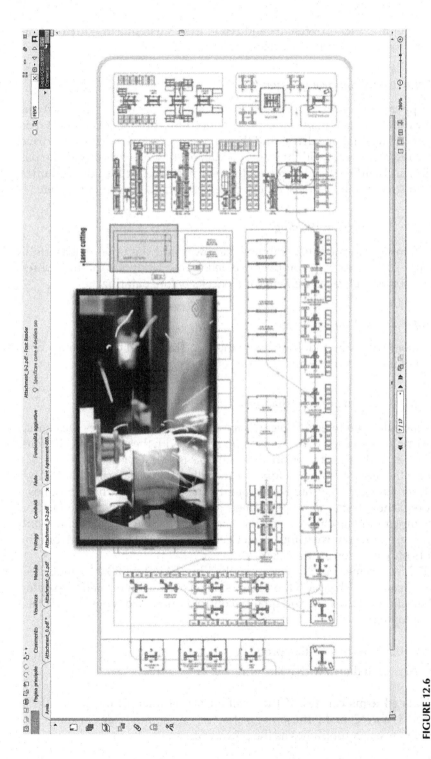

FIGURE 12.6

CNC automated laser-cutting of chassis tubulars. (From IFEVS 2018.)

passenger and pickup versions and ready to be customized with a specific upper subassembly frame according to its designation.

12.5.2.6 Upper Sub-frame Assembly

The upper subframe assembly is fabricating the upper part of the car frame. Few variants of upper subframe are produced for different car models (passenger car, hosting one driver and three passengers; taxi, hosting one driver and three passengers with luggage; pickup, hosting one driver and one passenger, and will reserve the rest of the rear part free for load or customization for specific applications, for example, restaurant, goods delivery, etc.

12.5.2.7 Framing Geo Welding

The framing geo module combines the lower and the upper subassembly frames. The output is a complete welded frame ready for the final three-dimensional (3-D) measurements and checks before painting.

12.5.2.8 Door Welding

The doors welding module delivers the two structures of the doors, composed of the outer frame and the anti-intrusion bar.

12.5.3 3-D Testing of the Chassis

This module is an intermediate quality-control area. It validates the chassis, measuring its geometry and comparing results with the expected values and tolerances. If the result is positive, the chassis is ready for painting.

Since the assembly and welding of the subpart can be carried out only if the laser-cut steel bars interlock properly, the geometry of the subframes is ensured; thus, tests can be performed only statistically given the fact that the templates themselves do not allow mistaken positioning of items—only in the case of heavy wear the correct geometry cannot be ensured, but this occurs only over long timescales.

3-D Measurement technology (supplied by (FARO AGBS 2018)) has been chosen to test the chassis of the vehicles: Thus avoiding most of the operators and machines downtime and quality-control issues. Two types of tools have been adopted:

- Articulated arm with a probe tip.
- Articulated arm with a laser line optical scanner.

Articulated arms can test if the position of the interlocking points, to be welded, is correct and if the joint position among sub-modules matches with the tolerances on subframes as the powertrain one.

To perform tests on the geometry of the whole chassis, articulated arms with 3-D optical scanners allow to collect large amounts of data and characterize parts in detail; critical surfaces and surroundings can be measured with high levels of confidence and speed. This technique represents an effective way to carry out Non-Contact Inspections on the chassis shape and allows to perform point-cloud-to-Computer-Aided Design (CAD) comparisons; therefore, any deviations from nominal geometry can be immediately detected.

12.5.4 Chassis, PWT Frame, and Door Painting Area

The painting area oversees painting and giving the proper protection to the complete chassis between the two powertrain frames and the doors structures. This is the last step before starting the final assembly of the car. Painting is done by deep immersion (preferred) or by robots.

12.5.5 PWT Assembly

The PWT assembly module produces the front and the rear axles. Motor, drive, gearbox, shock absorbers, brakes, and wheel hubs are assembled on the painted PWT frames.

12.5.6 Axle System Tests

Motor and inverter have to be paired; several figures of merit, related to the type of electric motor and its integrated position sensor, have to be set up. With the aid of a software graphical interface the parameters, such as nominal power, maximum spinning velocity, input voltage, and maximum current are set in the inverter firmware. Then the test verifies if the inverter properly drives the electric motor (forward and reverse spinning, position reading, acceleration and speed control, current absorption, etc.).

12.5.7 Composite Interior Panels

One of the first phases of the inner chassis assembly is mounting of the plastic interior panels. These components seal the inner car body, protecting from water, dust, and atmospheric agents.

12.5.8 Battery Box and Electric and Electronic Device Assembly

Once the mechanical part of the vehicle is ready, the battery box is added, together with the electrical and electronic (E-E) items and connections: All the control boards and the wiring are installed. IFEVS electric cars feature several identical control boards, each one dedicated to a specific group of services (traction, lights, mechanical actuators, climatization, infotainment, etc.); this allows to distribute the computational operations and reduces the complexity of the system.

12.5.9 PWT Testing Area

The overall performance of the powertrain is tested with a 4WD Chassis dynamometer. It consists of a rolling test-bench by BAPRO (BAPRO—Chassis Dynamometers 2018) (see Figure 12.7) which fulfills the stringent requirements to test high-power heavy vehicles as well. The system also includes a software suite to acquire and save measurement data and compare results. A remote control allows to freely manage the system without stopping the vehicle.

This 4WD platform can test the whole vehicle propulsion and transmission system at once.

Measurements of the main figures of merit (e.g., torque, shaft power, wheel power, wasted power) of the whole propulsion chain—including the inverter, the motor, and the transmission—can be evaluated.

The following tests are performed:

- Power sweep test.
- Cooling system check.
- Vehicle tests under constant speed/torque load.

The last option simulates driving on roads with different slopes or with different loads.

12.5.10 Body Exterior Panels

One of the latest phases of the vehicle assembly is the installation of the exterior panels. In this stage the car takes its final look.

FIGURE 12.7
The BAPRO dynamometer system used to test electric vehicles.

12.6 Factory System Architecture

A top-down approach has been applied to identify the relevant requirements and their relation to the Key Performance Indicators (KPIs).

The platform consists of a Manufacturing Execution System (MES) that contributes to the following tasks:

- Production scheduling.
- Traceability of production.
- Traceability of used materials.
- Control of the production flow.
- Visualization of the Key Performance Indicators (KPIs).

The framework used to identify requirements and KPIs follows the schema shown in Figure 12.8.

Not the least, human roles have been carefully considered, and associated requirements have been identified for the MES user interface (UI) and the human-machine interface (HMI).

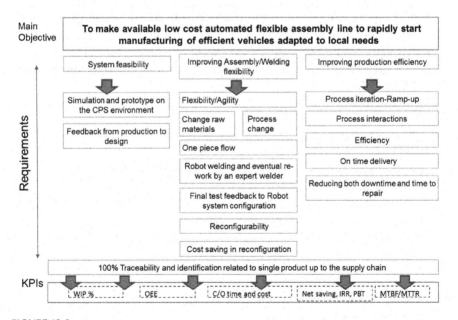

FIGURE 12.8
Framework to identify requirements and KPIs. (From EU HORIZON 2020 FoF PERFoRM Deliverable D1.4 2016.)

TABLE 12.1

Human-in-the-Mesh Requirements on Human-Machine Interface (HMI)

Context-aware mobile devices (role, location)
Intuitive representation of alternatives and trade-offs
Decision support enhanced by experts' decision-making patterns

Source: EU HORIZON 2020 FoF PERFoRM Deliverable D1.4 (2016).

12.6.1 Human Interfaces

The analysis of possible scenarios for human integration in the flexible production systems within these PERFoRM industrial use cases has been undertaken. Two types of human roles have been identified: Human-in-the-Mesh (see Table 12.1) and Human-in-the-Loop (see Table 12.2).

The final architecture is based on the following specifications:

- MES: Processes orders (submitted manually or on the Web), monitors, and visualizes KPIs.
- Scheduler: Plans services to optimize the production.
- Working islands: Feature part identification sensors, feedback lights, and HMI.
- Testing area: Tests are performed on sub- and complete systems, results are fed to the MES.
- Based on COMAU analysis on the new chassis, the factory system architecture has been optimized and simplified as shown in Figure 12.9.

TABLE 12.2

Human-in-the-Loop Requirements on Human-Machine Interface (HMI)

Context-aware mobile devices (role, location)
Support visual inspection with sensors
Support testing (geometrical, power train, fatigue, etc.)
Virtual presence (for consulting expert colleagues: sharing view, screen, info, voice connection, or chat)
Multimodal interaction (voice, image, gesture recognition, sounds, lights, etc.) to alert and support fieldwork
Suitable/wearable device to support fieldwork
Asset tracking (tools and spare parts)
Localization and turn-by-turn navigation to retrieve machines, tools, and spare parts.

Source: EU HORIZON 2020 FoF PERFoRM Deliverable D1.4 (2016).

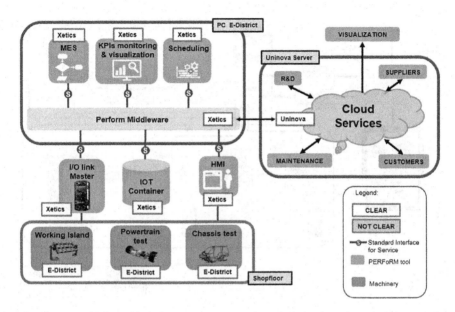

FIGURE 12.9
Factory system architecture.

IO-Link is a powerful standard, an increasingly deployed point-to-point serial communication protocol used to communicate with sensors and/or actuators. Extending the globally recognized Programmable Logic Controller (PLC) standard IEC 61131, it allows three types of data to be exchanged—process data, service data, and events. The I/O link Master has already an Open Platform Communications – Unified Architecture (OPC-UA) server onboard and can communicate directly with the Middleware developed by Xetics (XETICS GmbH 2018). The I/O link Master monitors all the sensor of the working island. The HMI is based on XETICS LEAN as well and runs on an Android-based handheld device. Every island has a dedicated HMI.

In Figure 12.10 the workflow of the demonstrator is shown. It follows the following steps:

- Customers order vehicles on Web apps.
- Orders are received by MES.
- MES sends instructions to operators via handheld devices.
- The operators follow the instructions to mount the steel tubes on the templates.
- The sensors identify components.
- The operators weld the parts.
- Upon completion operators return a feedback about the outcome.
- Results are fed to MES through the HMI.

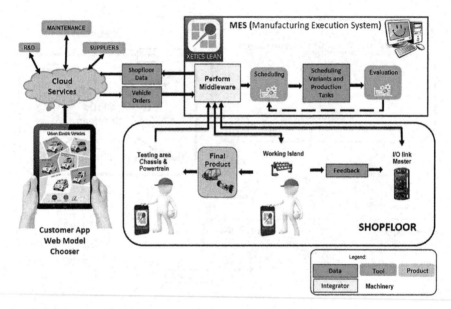

FIGURE 12.10
Workflow of the demonstrator.

- The products are verified in the chassis and powertrain test area.
- Factory data can be accessed through cloud services for analysis.

Xetics developed MES that includes the scheduler and KPI monitoring and visualization tools, and the HMI on android devices; Uninova (Instituto de Desenvolvimento de Novas Tecnologias, Portugal 2018) provided the Web app and cloud services.

12.7 Production Planning

Within XETICS LEAN, the Planning Board supports the manual scheduling and the production planning by optimizing worker and machine allocation. The production planning includes the following features:

1. Production tasks forwarding to available resource at the earliest time based on an ordered list.
2. Production scheduling to easily change the manufacturing sequence.
3. Resource assignment variations while assuring that only allowed changes are made.
4. Synchronization with the latest progress on the equipment.

The planning view is split in three sections:

1. Planning panel, showing resource allocation over time.
2. Planning list, containing all orders that are planned and visible in the planning panel. The order of this list is the base for dispatching the steps on the machines.
3. Overview list, containing all unplanned orders. Filters and sorting help the planner to focus on the relevant entries.

12.7.1 MES

XETICS LEAN is used to manage and monitor work in progress on the factory floor. It keeps track of all manufacturing information in real time, receiving data from machine monitors and employees. Although MES used to operate as self-contained systems, XETICS LEAN integrates an enterprise resource planning (ERP) suite, as shown in Figure 12.11. By improving productivity and reducing cycle- and total times to process orders, factory managers can ensure delivery of quality products in timely, cost-effective ways.

12.7.2 HMI

The HMI, running on handheld devices, introduces a higher level of automation:

1. Production tasks are submitted to specific operators.
2. Workpieces are placed on the assembly template.
3. HMI and Smart-lights signal incoming tasks (IN).

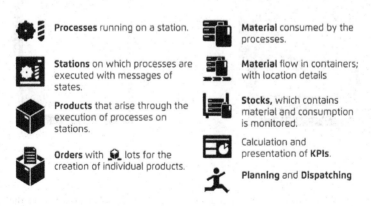

Processes running on a station.

Stations on which processes are executed with messages of states.

Products that arise through the execution of processes on stations.

Orders with lots for the creation of individual products.

Material consumed by the processes.

Material flow in containers; with location details

Stocks, which contains material and consumption is monitored.

Calculation and presentation of **KPIs**.

Planning and **Dispatching**

FIGURE 12.11
XETICS LEAN menu options.

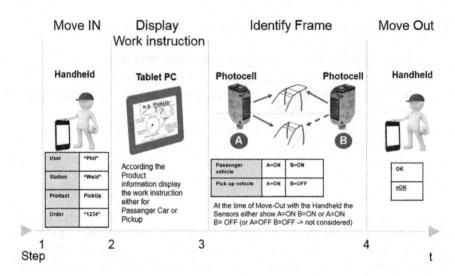

FIGURE 12.12
Example of the roof working island.

4. Via HMI, operators can follow the guided procedure (example given in Figure 12.12):

   a. Acknowledge the task (see in Figure 12.13, the Move IN step)

   b. Follow instructions for item selection and positioning

   c. Receive blocking warning (also on Smart-lights) in case of wrong operations

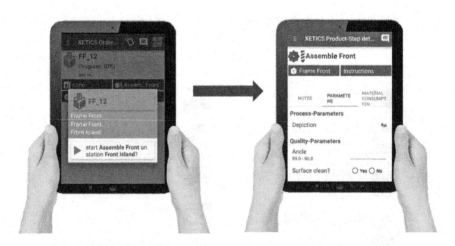

FIGURE 12.13
HMI Graphic User Interface—the operator acknowledges a task scheduled on the specific working island (left); upon completion the operator enters production and quality parameters (right).

    d. Upon completion, enter quality parameters and number of worked-out parts and scraps

    e. Confirm task accomplishment (see in Figure 12.13, the Move OUT step)

5. Workpieces are moved to the next station.

6. HMI and Smart-lights signal task completion (OUT).

An example of the guided procedure provided in the HMI is shown in Figure 12.14.

12.7.3 Traceability of Production

Tracking production steps on a given equipment retrieves all associated operation and quality parameters, as shown in Figure 12.15.

Material containers, as well as single subcomponents, are traced with QR/bar codes by android portable devices running the HMI (see Figure 12.16).

12.7.4 Storage of Cloud-Based Data

The XETICS LEAN is cloud-based; it can be accessed via Web user interface (Web-UI). The allocated relational Data Base (DB) System is also cloud-based. The supporting Middleware (Equipment integration) can be located either in the cloud or on premises (recommended).

12.8 Control of the Production Flow

In the overview tab of the XETICS LEAN dashboard, there is a widget containing the status and progress of the production orders and related jobs.

12.8.1 Timeline

The state distribution diagram of equipment shows how the conditions (STAND-BY, REPAIR, PRODUCTIVE, etc.) are distributed within a given period.

XETICS LEAN provides status details of each working island at any time; an example of the graphic user interface is given in Figure 12.17.

12.8.2 Analysis of Product and Process Deviations

The continuously updated analysis of all tracking events for stations, products, and processes retrieves start and end times as well as total durations of all product steps.

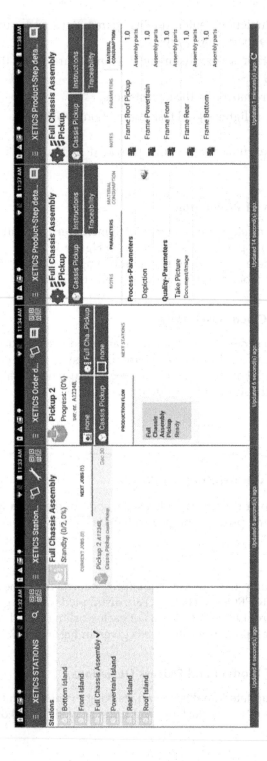

FIGURE 12.14
XETICS LEAN HMI down to order.

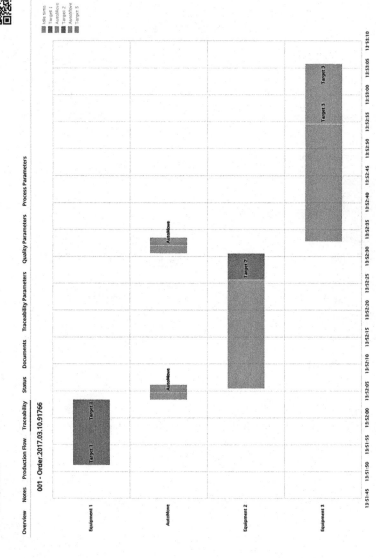

FIGURE 12.15
XETICS LEAN traceability tool.

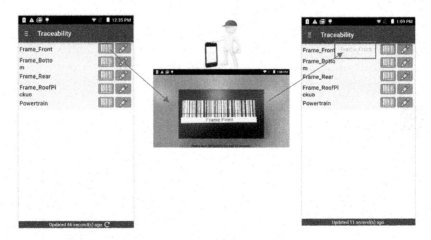

FIGURE 12.16
QR/bar code tagging for traceability.

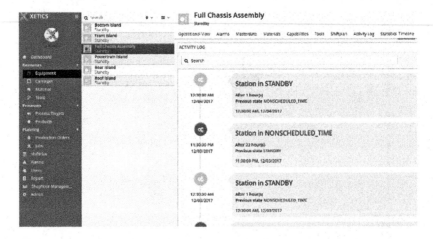

FIGURE 12.17
The XETICS LEAN interface showing the status of a given working island along the timeline.

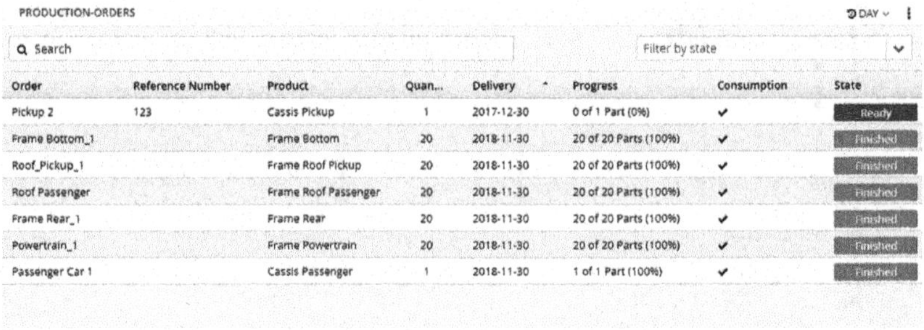

FIGURE 12.18
The XETICS LEAN production analysis tool.

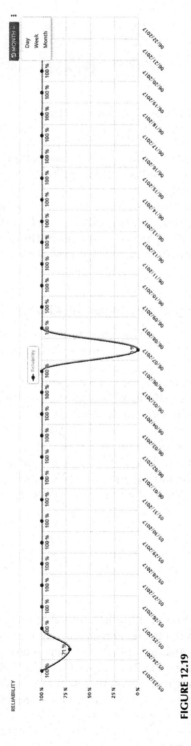

FIGURE 12.19
XETICS LEAN deviation analysis.

FIGURE 12.20
XETICS LEAN continuous KPIs update and monitoring tool.

Figure 12.18 provides an example of the production monitoring tool; for each scheduled task the scheduled delivery time is given, as well as the progress index and the status.

Minimum and maximum times are shown, as well as the deviations from allowed maxima; thus, it is possible to sort the table by deviations, filter by station, product, or process to find bottlenecks in the past production efficiency (Figure 12.19 exemplifies such analysis over a time period).

12.8.3 Visualization and Monitoring of the KPIs

The OEE and its factors (e.g., availability, utilization, throughput, and quality KPIs) are continuously updated. Figure 12.20 illustrates how the KPIs are provided in XETICS LEAN.

The IO-Link protocol, the IO-Link Master, Sensors, and Smart-lights have been successfully tested, including the OPC-UA server and XETICS LEAN communication, for fast, flexible, agile, and efficient production.

The cloud environment, implemented by Uninova, brings the new concept of cloud manufacturing to the electric car industry. The idea is a cloud environment, deployed into a dedicated server with a Web interface; it can store relevant data collected on the shop floor and share this data with external Web-based applications that can access this information at any time and place.

12.9 Conclusion

The application of the PERFoRM approach aims at achieving the next generation of digitalized, harmonized, flexible, and dynamically reconfigurable industrial manufacturing systems, enabling agility, evolution, self-organization, and adaptation along the system life cycle. This impacts facing the challenges of continuously and rapidly changing market conditions and increasingly smaller lot sizes, shorter lead-time, and time-to-market requirements. Modular plug-and-produce components (digitalized with built-in intelligence) are the basis of the IFEVS use case. All the involved actors (module suppliers, system integrators, end users, etc.) are then integrated to smoothly design, deploy, ramp-up, operate, and reconfigure the new generation of electric vehicle production systems, as it has been exemplarily done with the Microfactory of IFEVS.

According to the PERFoRM aims, the existing barriers have been overcome, and research and development results have been implemented into reference architectures while supporting Agile Manufacturing Control System for true plug-and-produce devices and human-machine integration and automation.

In these terms, methodologies and technologies have been developed and implemented to allow the production of different EV architectures aiming at establishing the conditions for a near future smart and flexible manufacturing of Urban EVs.

Methodologies and technologies specifically include:

- addressing quick adaptation of the production line from small lots of specific vehicles to larger lots of passenger cars;
- minimizing human errors when selecting settings and parameters;
- minimizing the variability of operations such as manual welding of the tubular chassis by introducing a high degree of robotized laser-cut and assembling and welding templates;
- satisfying the highest automotive quality standards by tracking the incoming parts, sub-modules processes, and integration phases;
- introducing testing methodologies on sub-modules before their integration;
- introducing testing and quality evaluation methodologies on final products before delivery;
- setting a database for updates; and
- providing a safe, ergonomic, and pleasant living environment.

The production steps have been simplified and optimized to be compatible with the flexible and agile production line, capable of giving an output of 50 vehicles per day, as described by COMAU based on the original study made by Torino E-District with Magneto Automotive: Seven islands featuring complex templates with sensors and feedbacks for real-time automatic control of all production operations.

The cloud services and the highly automatized control software of the proposed Microfactory have been successfully implemented by UNINOVA and XETICS, respectively: The whole system can handle and assist the management of all required operations from the client order on the Web to the final product delivery.

References

ACEA. 2018. "European Automobile Manufacturers Association." Accessed October 11, 2018. https://www.acea.be/.

Automotive Manufacturing Solutions AMS. 2015. Accessed October 16, 2018. http://www.automotivemanufacturingsolutions.com/focus/alliance-in-action.

Automotive News—Europe. 2015. Accessed October 16, 2018. http://europe.autonews.com/article/20150305/ANE/150309902/fiat-chryslers-polish-plant-will-build-restyled-fiat-500-lancia.

BAPRO—Chassis Dynamometers. 2018. Accessed October 18, 2018. https://www.bapro.it/de/.

Bloomber-1. 2016. Accessed October 16, 2018. https://www.bloomberg.com/news/articles/2016-01-04/gm-invests-500-million-in-lyft-to-bolster-alliance-against-uber.

Bloomber-2. 2016. Accessed October 16, 2018. https://www.bloomberg.com/news/articles/2016-08-01/daimler-boosts-blacklane-stake-as-ride-sharing-market-heats-up.

Business Insider UK. 2016. Accessed October 16, 2018. http://uk.businessinsider.com/tesla-driverless-ridesharing-plans-could-take-on-uber-2016-10?r=US&IR=T.

CARS, everything electric. 2010. Accessed October 16, 2018. http://www.cars21.com/news/view/459.

COMAU Spa. 2018. Accessed October 12, 2018. https://www.comau.com/EN.

Dailynews. 2018. Accessed October 10, 2018. http://www.nydailynews.com/opinion/turns-uber-clogging-streets-article-1.2981765.

EU HORIZON 2020 FoF PERFoRM Deliverable D1.4. 2016. Unpublished manuscript, last modified October 15, 2018. http://www.horizon2020-perform.eu/index.php?action=documents.

EuroNCAP. 2018. "The European New Car Assessment Programme." Accessed October 15, 2018. https://www.euroncap.com/de.

FARO AGBS. 2018. Accessed October 15, 2018. https://www.faro.com/de-de/produkte/bausektor-bim-cim/faro-focus/.

IDTechEx. 2011. "Idtex, New Revelations at Future of Electric Vehicles." Accessed October 16, 2018. https://www.idtechex.com/journal/print-articles.asp?articleids=2898.

IFEVS. 2018. "Interactive Fully Electrical Vehicles SRL." Accessed October 12, 2018. http://www.ifevs.com/.

Instituto de Desenvolvimento de Novas Tecnologias, Portugal. 2018. "UNINOVA." Accessed October 16, 2018. https://www.uninova.pt/.

RAND Corporation. 2018. Accessed October 16, 2018. https://www.rand.org/topics/autonomous-vehicles.html.

TESLARATI. 2016. Accessed October 16, 2018. https://www.teslarati.com/bmw-self-driving-cars-ride-sharing-service/.

XETICS GmbH, Germany. 2018. "XETICS LEAN Smart Factory." Accessed October 15, 2018. https://www.xetics.com/.

13

PERFoRM Approach: GKN Use Case

Per Woxenius, Stefan Forsman, Krister Floodh
(GKN Aerospace Sweden AB)

**Waldemar Borsych, Jeffrey Wermann,
Armando W. Colombo**
(Institute for Informatics, Automation and Robotics (I'AR))

CONTENTS

13.1 Introduction

The company GKN (Company GKN 2018), partner of the EU HORIZON2020 FoF Production harmonizEd Reconfiguration of Flexible Robots and Machinery (PERFoRM) project (EU HORIZON2020 FoF PERFoRM 2015-2018), has developed a novel flexible automation cell together with leading European Universities and Research Institutes, following the PERFoRM approach. The installations include a robot, mechanical installations for docking system and fences, electric installations, safety systems, and process modules. A substantial part of work has also been the development of robot paths, programming of robot and Programmable Logic Controller (PLC), as well as integration of reconfiguration functions.

The overall goal for GKN and the Use Case in applying the PERFoRM approach has been to develop industrial systems that increase automation, digitalization, flexibility, and reconfigurability to improve productivity, reduce lead times, and improve resource utilization.

An initial idea and principle description of a modular production concept "Micro-Flow Cell" was developed at an early stage (see Figure 13.5). For the Use Case, the main goal has been to develop and demonstrate the necessary solutions for simple and quick changeovers in the modular system, that is, an application of the principles for Industrial Digitalization and Cyber-Physical Production Systems (CPPSs). For this purpose, mechanisms for the seamless reconfiguration of the production process have been addressed and a demonstrator cell has been designed and documented in EU HORIZON 2020 FoF PERFoRM Deliverable D10.2 (2017). The solution includes several process modules, and the sequence of "shut down → plug out → replace module → plug in → start up" was tested and evaluated in the PERFoRM Use Case.

13.1.1 Summary of Scope of the System Goals and Requirements

As stated in EU HORIZON 2020 FoF PERFoRM Deliverable D10.1 (2016), the overall goals and requirements of the Use Case are as follows:

- A flexible, reconfigurable, and production cell concept, adaptable to different applications.

- Reduced time for changeovers and adaptation to current demands.
- Increased resource utilization.
- Reduced lead times through flow integration.
- Increased level of automation and digitalization for implemented processes.

13.1.2 Flexibility and Reconfigurability Definitions

To avoid any misunderstandings, definitions of Flexibility and Reconfigurability used in the GKN Use Case have been discussed in the upcoming sections.

13.1.2.1 Flexibility

The definition of and requirements for flexibility are listed below. The functions in the cell must be adapted to manage these requirements, but this is mainly accomplished through the design of the individual modules and the control system:

a. Product and process flexibility is required to provide the ability to make changes to the production equipment to produce different product variants or product types in the same cell, and to adapt to varying customer demands and to changes when new parts are introduced.

b. A change is required in the mix of current products produced to be able to handle the situation of planning and scheduling and possible changes in workload and capacity as well as use of tools and other process materials or consumables.

c. Introduction of a new product variant or type to be able to have flexible tooling and fixtures, or have the ability for quick changeover on the process modules. There is also an option to change process modules in the cell. However, this is more of a function of reconfigurability

d. The flexibility of the process equipment and its control is required to be able to make a variety of operations and geometries/features on the product using flexible tools or tool changers.

13.1.2.2 Reconfigurability

The goal with the function to do reconfiguration is to provide more powerful means to adapt to different situations and larger changeovers in the system and change the functions in the cell. As a lab and test bed, it will also allow easier access and availability for different users. Types of changes and recon-figurations are as follows:

a. Change process modules in the cell—that is, plug-out and plug-in another process module.

b. Modify or rebuild the cell and/or process module to another pro-cess function—that is, a large changeover that cannot be managed within the boundaries of flexibility.

c. Introduce a new type of process module—that is, a new process module for another type of process, or for a product that does not fit any current process module.

d. Move the process module to another cell—that is, the function can be developed at one location and moved, or shared between differ-ent value streams.

Any of these functions for reconfigurability is expected to be done with a "plug-and-play" functionality, which means that all process planning, pro-gramming, testing, and qualification is done and available in the cell, which is just configured to operate the current setup.

13.2 Scenarios and Work Processes for the "Micro-Flow Cell" Concept

In this section, first a holistic view of how the Micro-flow cell concept can be used in a future industrial system is described, followed by the more specific description of the workflow for how to operate the cell system and make changes.

13.2.1 Scenario for the Use of "Micro-Flow Cell" Concept in an Industrial Context

The long-term vision and goal with the technologies and results from the GKN use case is to contribute with solutions that enable an industrial sys-tem with higher degree of flexibility and reconfigurability of automated or semiautomated discrete manufacturing cells. The cell system should be a platform solution that can have different configuration of produc-tion processes. The processes can work individually or in an operation sequence, as the "micro-flow cell" concept defined in EU HORIZON 2020 FoF PERFoRM Deliverable D10.1 (2016). The cell scheduling/operation sequencing and control, integration, and reconfiguration in a heterogeneous system are the set of key functions/tools ("Cell Middleware"). For both long-term planning and near real-time planning and scheduling, the business planning systems need better integration than what is being used at the present time. Thus, the vertical integration with the business-level systems to automate data/information management for the production execution and reporting results ("Factory Middleware") is essential in the long-term perspective.

Outlined in Figure 13.1 is a holistic view of the information flow and activities for planning, development, and operation. The long-term planning and development of required solutions will be just as important as the short-term perspective on operations and execution to get the best possible effect of the advantages of the PERFoRM concepts and solutions.

The upper part of Figure 13.1 describes how the long-term planning and development of production planning as well as how the industrial structure and system solutions need to be done in an integrated way and plan for the use of the new cell concept. The main activities with numbered reference to Figure 13.1 are as follows:

1a. The long-term analysis of sales/demands gives answers to the need of production capacity, and which kind of capacity, using Rough Cut Capacity Planning (RCCP). Different scenarios can be simulated and analyzed (based on available/wish for resources). The result is an "Operation and Production Plan" (OPP) or "Master Production Schedule" (MPS).

1b. With input from the Sales and Operations Planning, there may be needs to adapt the functions and/or capability or capacity of the production system and its resources. This also includes the need of and advantages from having flexible and reconfigurable equipment/resources. The decision can be to acquire more resources, that is, some long-term technology development and investments, etc. as well as competence and skills of the workforce. It can also be an option of using/specifying need of flexibility and reuse/reconfigure equipment or parts of the production system. The result is technical requirements/specifications of the required production resources and plan to implement.

The lower part of Figure 13.1 describes the short-term perspective for production operations in which the cell concept is used.

2a. The "Material Requirement Planning" (MRP) and definition of available resources are inputs to make a more refined and detailed plan. This can be supported by simulation (however, currently not used) or other planning tools. The result is an MRP for material supply and Production Scheduling (PS) for the value streams.

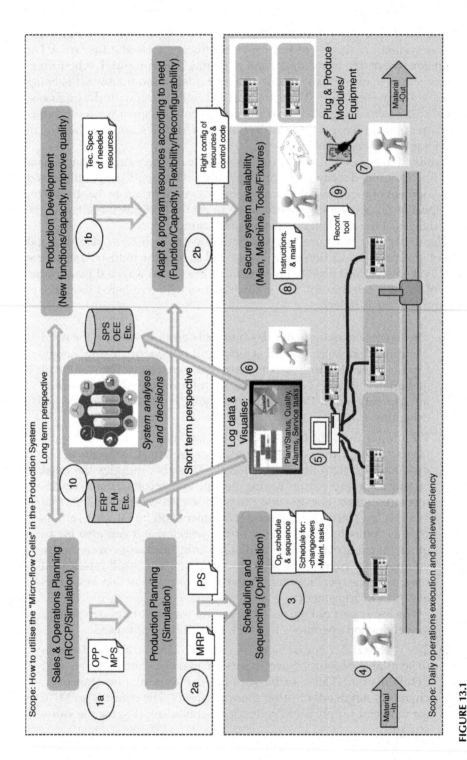

FIGURE 13.1

An outline of the future information and data flow structure from business systems to shop floor nodes.

2b. The production resources need to be prepared and set up with the correct functionality and configuration. Depending on the level of flexibility etc., much of the equipment can be reused and adapted to the current demands. This can include rebuilding a cell, rebuilding/ getting new tools, reprogramming, etc. The result is a correct configuration of work instructions, programs, HW equipment, etc.

3. With input from the "Production Schedule," the daily scheduling and sequencing of jobs at the "cell-level" should be as optimal as possible—that is, using the PERFoRM tool, for example, with a GKN SOLV tool, which has been investigated and defined in EU HORIZON 2020 FoF PERFoRM Deliverable D2.3 (2016). The result is an optimized schedule and sequence for the next shift or "X" hours. This plan can also create the slots and schedule a time for the changeovers, that is, when to change from one process module to another, as well as take into account the need for maintenance and create slots for that on the schedule.

4. The jobs are triggered by the system/schedule. However, start and execution is not done 100% automatically, as the operator's first need to load/feed required material into the cell, and start the execution. The operator may also have different tasks, as a shared resource, and perform tasks from different schedules.

5. The cell controller executes the automated activities of the production schedule, when the operator releases the jobs.

6. The operators are supported by supervision and visualization from the system to show status of jobs/orders, equipment, etc. and display on the Monitor or Human-Machine Interface (HMI). Local real-time and history databases in the cell can be used. Some part of that data is also sent to lower- or upper-level systems/database(s) for further use: for example, the PERFoRM "tools."

7. The operator unloads parts from the cell, and after mandatory inspection and confirmation of quality is done, it finalizes each job to report the completion.

8. Depending on the type and level of designed and available flexibility, there are additional tools, fixtures, and process modules for changeovers in the cell—that is, "Plug-and-Produce" equipment "ready-to-use"! This requires routines, plans, and instructions; for example, maintenance, standard work, etc. to use the production resources.

9. To execute the "Plug and Produce," a "Reconfiguration Tool" in the cell control system is required to manage the changeover process.

10. The data made available from the production cell/system provides the source to visualization of Key Performance Indicators (KPIs), to be used in different "tools," analyses, and decision support for improvements and planning, scheduling, etc.

The industrial structure needs to be developed and constructed step by step over a long period of time, and following the needs for replacement of old equipment and upgrading technology and/or production capacity. The information system at the top level should be implemented together with the first applications of the automated cells, which will then facilitate and shorten the lead time for introduction of new "micro-flow" cells. As illustrated in Figure 13.2, the production cell platform can be expanded and reconfigured for different applications and used to design parts of the production flow.

13.2.2 "Workflow" for Flexibility and Reconfigurability

For the GKN use case, as highlighted before, one of the main goals is to develop and demonstrate the necessary solutions for simple and quick changeovers in the modular system, that is, an application of the principles for Industrial Cyber-Physical Production Systems. The definitions and overall requirements for flexibility and reconfigurability are described in Section 13.3. In Figure 13.3, the role of humans, activities, and information flow for operating the cell is illustrated and the "workflow" summarized in the following paragraphs. The "workflow" for exchanging process modules in the cell is then explained and illustrated in Figure 13.4.

The generic operation sequence to use any of the processes in the cell and to execute a production order follows the main steps as described in this example:

- From the Enterprise Resource Planning (ERP) system, available orders are scheduled (1) and an optimized sequence is generated (2), managed by the Cell PC, and posted on HMI (3).
- The operator reads the available production orders and related information/instructions (4, 5) and prepare for each operation to make sure conditions are correct and all materials and tools at hand.
- Operator load "part" to magazine (6) and starts the job from the HMI (7).
- The system (Cell PC) checks that all requirements for work are fulfilled and sends tasks to be executed to the PLC (8).
- Robot is requested to start. The part ID is read/confirmed with the order number, and correct robot program is downloaded (9). This is controlled by the PLC.
- When the robot program is downloaded, the robot starts, getting the right gripper and load part in the fixture or tools, etc. (10).
- When the processing is finished, the robot returns the part to the magazine and the order status is set to done (11).
- The operations are repeated until all parts have been processed.
- Operator is informed that the order is processed (from the HMI), unload parts, check/inspect as needed and finalize the order and report in SAP Enterprise Ressource Planning (ERP) (12).

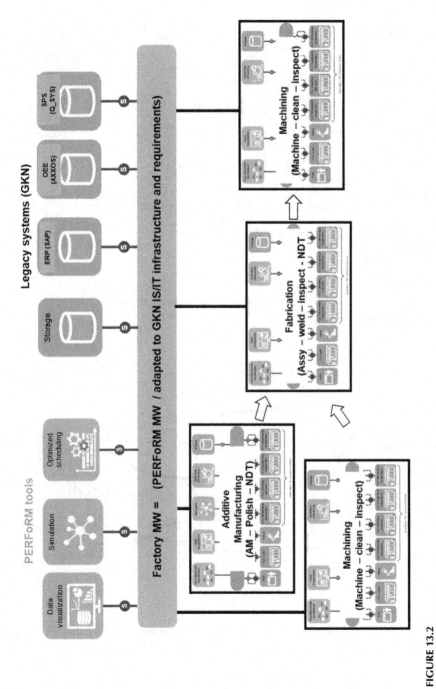

FIGURE 13.2
Illustration of a potential future modular and scalable architecture.

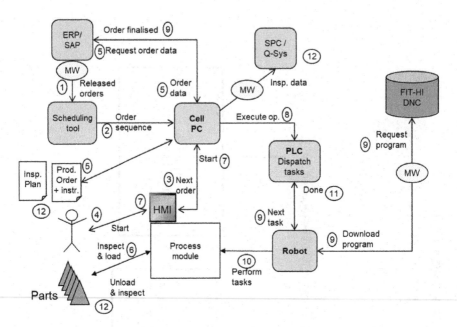

FIGURE 13.3
The "workflow" for operating the cell to plan and execute production orders.

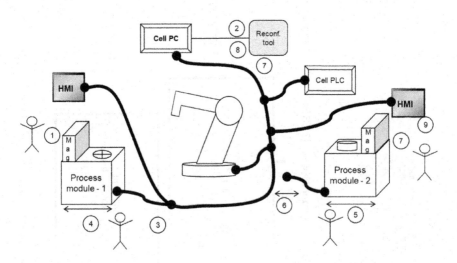

FIGURE 13.4
The "workflow" for planning and changeovers demonstrating the production flexibility and reconfiguration of the cell.

The cell system is designed and set up to handle different process modules. To test and demonstrate flexibility and reconfigurability, the sequence of *shut down → plug -out → replace module → plug in → start up* will be tested and evaluated. A generic procedure to replace any of the modules in the cell should follow the steps as in the following example:

- The operator's plan to stop production and start a sequence to prepare for a changeover is initiated on the HMI (1).

- The module is shut down and then "check-out" from the cell → managed by the "Reconfiguration tool" → communication through the HMI (2).

- The process module is (physically) disconnected—signal cables, media, etc. —and prepared to be moved (3). The process module is released and transported out of the cell (4).

- The other process module is transported into the cell, and positioned and locked, using clamping/positioning devices (5).

- The process module is prepared for production (any transport mode securings removed) and (physically) connected—signal cables, media, etc. (6)

- The process module is turned on, and performs a self-check to get ready to connect to the cell system (7).

- The identification of the process module is recognized by the "Reconfiguration tool," which starts the initiation and changes the settings in the cell controller (8).

- Confirmation is received on the HMI that the reconfiguration is done and the module is ready to be used (9).

13.3 Technical Requirements and Specifications

In this section, the functional description and requirements for the "Micro-Flow cell" concept and the use case demonstrator are defined. For the later design and implementation, it is expected that relevant standards and regulations are used and followed to ensure safety and security.

13.3.1 Cell Functions

The purpose of the cell is that it should be used for development and demonstration of different kinds of automation applications. The approach with a modular system will give the opportunity to build and implement different solutions over time. Specific functions and requirements will be relevant for each such process and need to be specified separately. In this section, for the cell system, the common main functions and requirements of the cell are as follows:

- Parts (product) should be manually loaded/unloaded into part magazine or fixture on a module.
- Loading of tools, process material, replacement of worn tools, etc. should also be done at defined storage positions/tool locations on the module.
- Visual signals on loading positions and/or information on HMI should be used to guide the operator to prevent misplacement, loading/picking of wrong parts, tools, etc. (Error Proofing).
- An automated process/operation should be done by a robot and the dedicated tooling (grippers, process tools, etc.) for a specific product should be stored on the product/process modules.
- The robot picks parts from magazines, loads parts to fixture, or holds the parts during processing.
- Inspection/confirmation of which part that is about to be processed should be done automatically, or with support from the system for the operator to confirm that everything is correct. The operator should get information, and if necessary give input, through an HMI.
- Manual or semi-manual operation, inspection, etc. should be possible if a process requires that, at least at some of the module stations.
- After finished processing, the robot returns parts to magazines, or move them to a fixture or a magazine on another module for the next sequence/operation.
- The general idea of the Cell is that it should be used for a sequence of operations, using two or several process modules. The robot can move the parts between the process modules (or the operator if needed) and execute the different operation sequences. However, it should also be possible to use each of the process applications individually, as a single operation on a product.
- The system needs to keep track of all parts, process material, tool usage/lifetime, etc. that need to fulfill the requirements for process monitoring, documentation, and traceability. The need for human supervision and intervention should be minimized. Specific functions and requirements will depend on each application/process, that is, they will be specified for each process module separately. Thus, the cell system must be prepared to handle such information, or be adapted to such functionality over time.

- A pallet/fixture change system (zero point system), etc. may be used on a process module to adapt to different product variants or part geometries. To prevent any errors or damages, the system must be able to detect which fixture/tool is in use.
- A process module docking system must be designed for easy, quick, and reliable changeover of different process modules. This includes both the mechanical interface for module positioning, as well as the media and communication connection(s) (see detailed requirements below).

13.3.2 Process Modules

For the different applications in the cell, the stations should be equipped with modules with a standardized interface for positioning and securing its location. In addition, the media and communication connection should use a standardized interface. This is required to fulfill the overall goals and requirements to build a system for plug-and-play.

For each application all the hardware, control, and software needed should be built on a module, an autonomous unit that, when plugged into the cell system, will be made available as a production asset, ready for production. Depending of the type of process and size of the product the application can use one or several module units to provide enough space. The design of the process and its module should allow them to be placed in different cell stations, that is, they should not be designed and dedicated to a specific slot. (Exceptions may be accepted due to environment or safety, if auxiliary equipment and/or encapsulation is needed.) Thus, it can be a single module that can be placed in any of the station, but it is also possible to build a two-unit module, or a three-unit module.

The following requirements for the module units should be met:

- The unit module chassis floor space should be (Length × Width) 1200 mm × 800 mm.
- Recommended height of the module chassis is 800 mm, to provide a convenient and ergonomic working position for the operator. However, what is built on top of the chassis can be very different depending on application.
- No parts or features can stick out from the perimeter of the chassis envelope, defined by the 1200 × 800 dimensions.
- A control cabinet should be assembled inside the chassis (with PLC, power management, pneumatic valves and distribution, etc.). Switches and connector locations must be easily accessible when the module is positioned in the cell station, and any light signals must be visible from operator position.

- The module should be transported with a forklift trolley, forklift truck, or integrated wheels on the module.
- The module should have a mechanical docking interface for positioning the module in the cell stations (the docking system will be the same in all stations). The tolerance of the docking interface and accuracy/stability of the module chassis should give a repeatability of the positioning of +/-1 mm.
- The connection of media (power, air) and communication should have a standardized interface (or interfaces). The connection should preferably be done at one point and with one connection only. Connection and disconnection must be simple and safe, and that it can be done by an ordinary operator.

13.3.3 Robot System and Tooling

13.3.3.1 Robot Functions/Requirements

- ABB IRB with minimum handling of 200 kg and minimum 2.5 m reach
- Force control function
- SafeMove 2
- Note: Preferably latest version of controller and software

13.3.3.2 Cabling/Hose Package Requirements

All cabling, both the external and robot mounted, must have a well-defined position and be designed so that mounting can be done only in one way, and provide the following functions to the tool changer.

- Ethernet-based bus
- Compressed air
- 24 V signal wiring
- 24 V power wiring

13.3.3.3 Tool Changer Requirements

- Tool changer must be adapted to the robot type and handling weights and transfer the media and communication to the tool.
- Tool changes will be controlled by the robot.
- Tool storage/change locations may be in the cell system as well as on the Process Modules.

13.3.4 Safety System

The safety system must be able to support the functions for flexibility and reconfigurability, but at the same time not compromise the safety and security of people, products, and the equipment. The following requirements should be fulfilled in a safe way:

- An operator must be able to access load/unload positions for parts, process media, and tools, without emergency/process stops.
- While the operator is attending/working in one module area, the cell/robot should continue its process/cycle on other stations/ modules.
- If the robot and operator are in the same module/station, the robot must stop. Alternatively using
  - activate a production stop, before entering the module area and
  - safety equipment and safe move.
- Unnecessary stops/emergency stops must be prevented. For example, if the robot is performing a task that should not be interrupted, there should be a signal to people, not to enter that zone.
- It should be possible to "plug out and plug in" (exchange) a process module in one station, while the robot is working in another module.

13.3.5 Integration, Communication and Media Supply

13.3.5.1 Integration and Communication

Based on the GKN use case architecture (EU HORIZON 2020 FoF PERFoRM Deliverable D10.1 2016), the following functions and requirements should be followed for design of integration and communication.

- Data communication Ethernet based using transmission control protocol (TCP)/internet protocol (IP) and Open Platform Communications Unified Architecture (OPC-UA) protocols. Wire as well as Wi-Fi may be used (preferably use the same connections/cables to each controller unit for programming, backups, diagnosis, etc. that are used for the cell production communication).
- 24 V signal wiring.
- Safe I/O signals should be available at all nodes/process modules, as well as for the safety system components when required. This also includes safety switches; for example, door/gate, emergency stops, etc.
- ProfiSafe is required for the ABB SafeMove 2 communication (and is an option for some of the Safety components).

13.3.5.2 Media

The cell will need different media for its functions, and required media supply to the cell, from the facility, is listed below. That media need to be distributed to each process module.

- Electrical power (400 V, Min 16 A).
- Electrical power, available on process modules (24 V, TBD A).
- Compressed air: (Min 6/Max 10 Bar, and at maximum flow TBD l/min).
- Water (or liquid media)—may be needed for some application, but will not be provided from the facility. A local solution for that process module must be used.
- Water drain—may be needed for some applications, but will not be provided from the facility. A local solution for that process module must be used.
- Exhaust/ventilation will be needed for polishing, grinding, etc. to remove dust (avoid pollutions in the cell), and it may also be required from safety and health perspectives. This will not be installed in the cell and connected to the facility system. Instead a local solution for that process module must be used, for example, an EX-classed vacuum cleaner.
- All cabling and routing of media inside and on the outside of the cell must have a well-defined position and connection points for the Process Modules. Any connections should be designed so that mounting can be done only in one way. This is a necessity for high availability, flexibility/reconfigurability, and offline programming.
- Note: The point of supply/connection has requirements but specified in the internal documentation of the company only

13.4 Implementation of Micro-Flow Cell Concept

This section describes the implementation of the selected technical concepts.

13.4.1 Cell Overview

Figure 13.5 depicts an overview picture of the Micro-Flow cell concept, implemented at the PTC, "Production Technology Center," in Trollhättan (EU HORIZON 2020 FoF PERFoRM Deliverable D10.2 2017).

The robot is located in the center of the cell and is able to operate all the stations and tool stands. There are five different stations. Three stations will allow an operator to assist each station and its modules to, for example, load/ unload parts, tools, process material, or other tasks needed for the respective

FIGURE 13.5
Installed "Micro-Flow Cell" for the GKN use case.

processes. It will also be possible to change modules in one station while the cell is running. Two stations are behind a stationary glass fence. Those stations/modules can also be replaced using the same plug-and-play function, but not while the cell is up and running. It must be stopped and shut down to remove the fence and make the changeover. This choice was made to keep the costs and complexity of the safety system to a reasonable level.

At the current stage, the installations of robot, electrical power, communication, module interfaces, and safety are completed.

13.4.2 Process Modules

A process module and the mechanical docking interface are depicted in Figures 13.6 and 13.7. The purpose of the module is to act as a component and tool storage for the spindle. In combination, the spindle and storage module can be used for brushing or deburring of vane components. The frame of the module is built using extruded aluminum beams and brackets.

13.4.3 Robot System and Tooling

The selected robot is an ABB 6700 model; Figure 13.8 shows a close view of the installed robot with the tool gripper attached. A pillar is used to mount the robot in order for the robot to be able to access all process modules. Tools

FIGURE 13.6
An assembled product module.

for the robot are stored on tool stands within the cell. The selected robot fulfills all functionality requirements except for the OPC-UA protocol; this functionality is instead achieved through the use of an adapter installed on the cell PC (EU HORIZON 2020 FoF PERFoRM Deliverable D6.6 2017).

13.4.4 Safety System

GKN has developed an advanced safety system to meet the requirements for how to use and access the cell. The chosen safety solution is utilizing the advantage of SafeMove 2 functions among monitor axis range, monitor standstill, or safe zones for the robot. The cell area was divided into five main zones (1, 2, 3, 4, and 5) corresponding to the five modules' blocks as it is described in Figure 13.9. The safety system uses a variety of components to monitor areas that the operator can access. Floor mounted scanners are used as the first step, to detect if any operator is approaching the outer safety

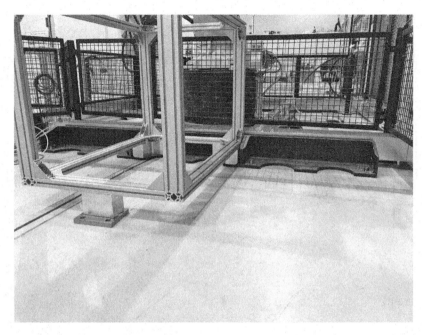

FIGURE 13.7
A module docked into a docking station.

FIGURE 13.8
Installed ABB 6700 robot.

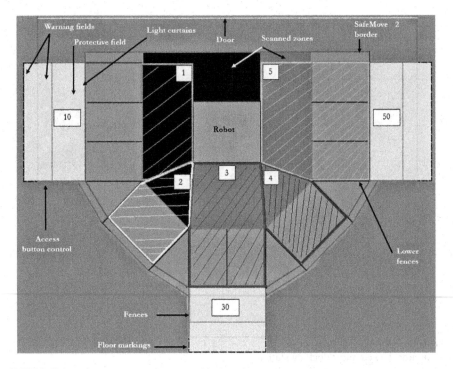

FIGURE 13.9
Drawing of the cell with safety zones and monitored fields in the module stations.

zones of the cell. Detection of an operator will limit the robot movement in the area where the operator is standing and limit the robot speed. The second step of cell security is the light curtains. Breach of a light curtain will stop robot movements entirely. Figure 13.10 depicts floor scanner and light curtain equipment installed in the cell.

13.4.5 Cell Communication

In this section a more refined description of the GKN Use Case architecture is briefly described, based on the results of innovation activities reported in EU HORIZON 2020 FoF PERFoRM Deliverable D2.2 (2016) and EU HORIZON 2020 FoF PERFoRM Deliverable D10.1 (2016). It is illustrated in Figure 13.11 and consists on a heterogeneous Industrial Cyber-Physical System using different ties and brands of PLCs, controllers, and HMIs, which are digitalized and networked using OPC-UA as communication and integration protocol (Gosewehr et al. 2017; Haskamp et al. 2017).

OPC-UA standard will be used as an integration layer that will enable the communication between the assets on the OT level and components and systems on the IT level. Biztalk technology, which is currently being used by GKN as a Middleware (Microsoft 2018), will be integrated with the OPC-UA

FIGURE 13.10
Light curtain and floor scanner making up the safety zones for the "micro-flow cell."

communication protocol. The necessary adaptation to those conditions and the application of the PERFoRM Modeling Language (*PERFoRMML)* and the PERFoRM Middleware component, reported by EU HORIZON 2020 FoF PERFoRM Deliverable D2.3 (2016) and EU HORIZON 2020 FoF PERFoRM Deliverable D2.4 (2017), were used for demonstration in the test bed (EU HORIZON 2020 FoF PERFoRM Deliverable D6.3 (2017); EU HORIZON 2020 FoF PERFoRM Deliverable D6.6 (2017)).

A necessary step for implementing the GKN system architecture was to find a concept and a practical solution for performing the connection between cell (Operational Technology (OT) digitalized using OPC-UA) and Information Technology (IT) components such as SAP ERP, using BizTalk message broker. That is, to define and test a solution for download of production order data and upload of the data generated at the production level. To find and confirm that the different kinds of data can be communicated and that the communication/data flow is possible in both directions (OT ⇔ IT). The full setup is shown in Figure 13.12, which did allow to demonstrate a successful communication between the cell-level PLC and the top-level SAP system.

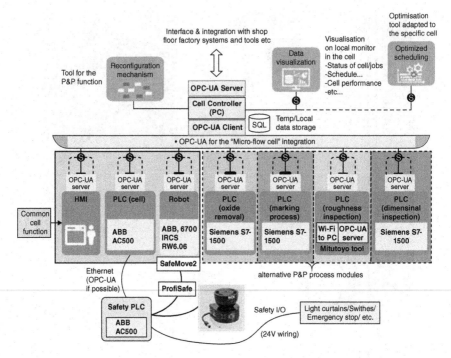

FIGURE 13.11
GKN's system architecture.

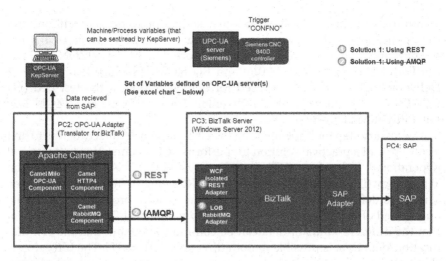

FIGURE 13.12
Workshop setup.

For one of the planned processes in the GKN use case, an adapter is required for substituting the wired serial communication of a Mitutoyo roughness sensor. The development and implementation of the sensor adapter hardware design and software development has practically successfully demonstrated at GKN premises. A detailed description of the adapter can be found in EU HORIZON 2020 FoF PERFoRM Deliverable D3.1 (2016). The prototype implementation of the Sensor Adapter is complete and working, and has been successfully demonstrated, as reported in EU HORIZON 2020 FoF PERFoRM Deliverable D6.4 (2017).

13.4.6 Cell Reconfiguration

To realize the reconfiguration functionality for the micro-flow cell, agents were implemented as described in the publication "Agent-based reconfiguration in a micro-flow production cell" (Dias et al. 2017). These agents are built upon Multi-Agent system (MAS) principles and OPC-UA communication for transfer of configuration information within the cell as well as information from shop floor to high-level production systems (MES, ERP). The agents were implemented and tested with an OPC-UA server between the cell and the agents, which contains all the necessary data points to enable the collaboration for both sides. As a result, great data flexibility remains on the part of the production system during implementation and neither the communication interfaces of the agents nor the robot cell hardware data point needs to be changed. Only the OPC-UA server for signal forwarding must be implemented when the agents are commissioned. In the following, this server is called the OPC-UA cell server. The interface of the OPC-UA cell server toward the agents, including two types of agents (robot and brushing module agents), is shown in Figure 13.13. On the other side, this server connects to the robot and the modules of the cell, which may have different IDs and namespaces for the data points. The OPC-UA cell server wires up these signals and forward production data to the agents and control signals from the agents toward the cell hardware.

The cell concept provides one robot for each cell, which cannot be plugged out. To establish the connection, the IP of this robot was assigned statically and the connection between the robot agent, the OPC-UA cell server, and the robot control unit is set up during commissioning of the cell and maintained for the entire work process. As soon as a module connects to the cell, the OPC-UA cell server needs to have information about the slot number on which the incoming module connects to the cell and its type to enable the previously described principle. To find out which slot the incoming module is connected to, a cell PLC with an OPC-UA interface is installed, which is monitoring the module slots of the cell using proximity sensors. For this purpose, an ABB AC500 PLC system (PM5650-2ETH) is used. The information about incoming modules is exposed via OPC-UA within the cell network. After the OPC-UA cell server receives a signal about an incoming module, the connection to this module needs to be established. To do so, first the IP-Address

FIGURE 13.13
An OPC-UA cell server agents interface.

of this module PLC needs to be obtained. To keep the flexibility within the cell, a Dynamic Host Configuration Protocol (DHCP) server, which provides addresses to connected modules and adds this information to a cell connections database, is installed on the Cell PC. After the OPC-UA cell server obtains information about an incoming module on a given slot, this database is queried and as soon as the IP address of the module is obtained, connection to the module can be established, and its type is read and forwarded to the agents. This completes the cell Plug-In process, and the new module is ready for use. Figure 13.14 illustrates the process with all components.

The Plug-Out process works in a similar sequence as Plug-In with the difference that the IP of an outgoing module is known, and therefore the database has not to be queried.

A disadvantage of this software setup is that before commissioning one module maximum should be connected to the cell. In the case when two or more modules are plugged in before starting up the cell software, the OPC-UA cell server will not be able to assign the module—slot relation. In further work, it is intended to implement a graphical user interface that opens in this case, so the operator can assign plugged in modules to slots manually.

The next step is to connect the scheduling modules and the PERFoRM Middleware component to the cell. The data flow from SAP ERP and Biztalk to the shop floor has already been tested. The biggest challenge in implementation

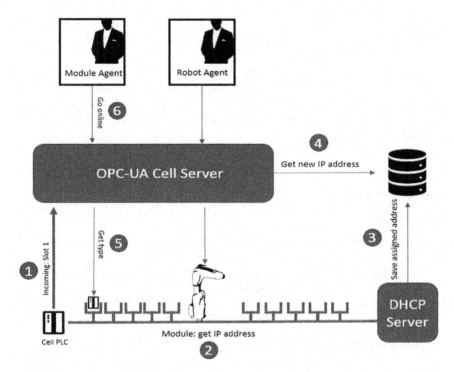

FIGURE 13.14
The Plug-In reconfiguration process.

of this concept in the cell is to construct a schedule schema that fits all needs for the cell as well as MES and ERP. Further, it is also important that the schedule is visualized and easy to understand for personnel involved in the process. The archetype of a schedule is planned to be constructed as XSD-file using PERFoRMML, with which each schedule automatically becomes an XML file with a known structure and can be easily interpreted by all components involved in the production process.

The visualization component shown in Figure 13.11 has been implemented by means of a KPI. Historical production data for each cell is stored in a database. The operator is then able to access the historical as well as the current production data and visualize it with a KPI.

13.5 Conclusion

This chapter summarizes the engineering steps and the major results of applying the PERFoRM approach to a "Micro-Flow cell" developed and installed at the GKN premises, in order to demonstrate how the PERFoRM methods and technologies support increasing automation, flexibility, and reconfigurability and consequently improving productivity, reducing lead times, and optimizing resource utilization.

For the use case, a major outcome has been the development, implementation, and demonstration of the necessary solutions for simple and quick changeovers in the industrial modular production system, that is, an application of the PERFoRM principles for this kind of "Industrial" Cyber-Physical Production Systems.

References

Company GKN. 2018. Accessed October 10, 2018. https://www.gknaerospace.com/.

Dias, Jose, Johan Vallhagen, José Barbosa, and Paulo Leitão. 2017. *"Agent-based reconfiguration in a micro-flow production cell."* Proc. of the 15th IEEE Int. Conf. on Industrial Informatics (INDIN2017), pp. 1123–1128. Emden, Germany.

EU HORIZON 2020 FoF PERFoRM. 2016. "Deliverable D2.2: Definition of the System Architecture." Unpublished manuscript, last modified 10, 2018. http://www. horizon2020-perform.eu/index.php?action=documents.

EU HORIZON 2020 FoF PERFoRM. 2016. "Deliverable D2.3: Specification of the Generic Interfaces for Machinery, Control Systems and Data Backbone." Unpublished manuscript, last modified 10, 2018. http://www.horizon2020-perform.eu/index.php?action=documents.

EU HORIZON 2020 FoF PERFoRM. 2016. "Deliverable D3.1: Adapters Implementation for at Least Three Different Production Resources." Unpublished manuscript, last modified 10, 2018. http://www.horizon2020-perform.eu/index. php?action=documents.

EU HORIZON 2020 FoF PERFoRM. 2016. "Deliverable D10.1: GKN Use Case Goals, Requirements and KPIs—Specification of Applications, Functions and Requirements for the "Micro-Flow cell"." Unpublished manuscript, last modified 10, 2018. http://www.horizon2020-perform.eu/index.php?action=documents.

EU HORIZON 2020 FoF PERFoRM. 2017. "Deliverable D2.4: Industrial Manufacturing Middleware: Specification, Prototype Implementation and Validation." Unpublished manuscript, last modified 10, 2018. http://www.horizon2020-perform.eu/index.php?action=documents.

EU HORIZON 2020 FoF PERFoRM. 2017. "Deliverable D6.3: Report on Self-Adaptive Machines Demonstrator Design and Set-up." Unpublished manuscript, last modified 10, 2018. http://www.horizon2020-perform.eu/index. php?action=documents.

EU HORIZON 2020 FoF PERFoRM. 2017. "Deliverable D6.4: Report on Self-Adaptive Machines Demonstrator Documentation and Results." Unpublished manuscript, last modified 10, 2018. http://www.horizon2020-perform.eu/index. php?action=documents.

EU HORIZON 2020 FoF PERFoRM. 2017. "Deliverable D6.6: Self-Adaptive Large Scale Demonstrator Documentation and Results." Unpublished manuscript, last modified 10, 2018. http://www.horizon2020-perform.eu/index.php?action=documents.

EU HORIZON 2020 FoF PERFoRM. 2017. "Deliverable D10.2: Concept Development of the "Micro-Flow Cell"—Development, Evaluation and Testing of Functions for the GKN Use Case." Unpublished manuscript, last modified 10, 2018. http://www.horizon2020-perform.eu/index.php?action=documents.

EU HORIZON2020 FoF PERFoRM 2015-2018. Accessed October 10, 2018. https://cordis.europa.eu/project/rcn/198360_de.html.

Gosewehr, Frederik, Jeffrey Wermann, Waldemar Borsych, and Armando W. Colombo. 2017. *"Specification and design of an industrial manufacturing middleware."* Proc. of the 15th IEEE Int. Conf. on Industrial Informatics (INDIN2017), pp. 1160–1166. Emden, Germany.

Haskamp, Hermann, Michael Meyer, Romina Mollmann, Florian Orth, and Armando W. Colombo. 2017. *"Benchmarking of existing OPC UA implementations for industrie 4.0-compliant digitalization solutions."* Proc. of the 15th IEEE Int. Conf. on Industrial Informatics (INDIN2017), pp 589–594. Emden, Germany.

Microsoft. 2018. "Automate processes with enterprise-wide application integration." Accessed October 10, 2018. https://www.microsoft.com/en-us/cloud-platform/biztalk.

14

PERFoRM Approach: Lessons Learned and New Challenges

Michael Gepp, Matthias Foehr

(Siemens AG)

Armando W. Colombo

(I²AR, University of Applied Sciences Emden/Leer)

CONTENTS

14.1 Introduction

The industrial environment experienced its first revolution with the transformation of manual work into steam-powered activities, in the period from about 1760 to sometime between 1820 and 1840. This transition included going from hand production methods to machines, new chemical manufacturing and iron production processes, improved efficiency of water power, the increasing use of steam power, and the development of machine tools. It also included the change from wood and other bio-fuels to coal. Few years later, ending the XIX Century, appeared the electrical machines replacing those steam-powered machines giving reason for speaking about a 2nd revolution on the shop floor. In less than 70 years of the XX Century, the apparition of the computer programs and its insertion in controllers like the Programmable Logic Controller (PLC) and the first robots started making a reality the initially automatized production environment, integrating computer-technology with control and automation systems, and this visible technological but also social transformation could be classified as the third industrial revolution.

Since the last decade of the 20th century, but more intensively during the first years of the third Millennium, the industrial environment has been witnessing rapid changes into an ecosystem with a high degree of multidisciplinary knowledge and intelligence embedded into devices and systems fully distributed across the company. The technological and organizational changes are mainly driven by business and societal needs that have been moving from traditional mass production toward mass and extreme customization of goods and the production systems working in the background. This evolution is supported by new disruptive advances both on software and hardware industries, as well as the cross-fertilization of concepts and the amalgamation of mechatronics, information, communication, and control technology-driven approaches in traditional industrial management, automation, and control systems.

In this industrial context, mechatronics, information, communication, and control systems combine progress achieved by the application of large distributed computing systems in product and production system life cycle with the power of digital data that is produced during manufacturing processes and also the data embedded in the products by many small data-driven devices, like sensors, actuators, or radio-frequency identification (RF-ID) devices. Moreover, these smart devices and systems, enabled by advances in machine to machine (M2M) communications and cognitive control systems, are connected, communicating and exchanging data and information in a networked manner, combining two major features: Autonomous decision-making processes and cooperation/collaboration capabilities to solve problems and approach situations at the same time, which can only be managed by the networked group and not by a unique component. All this confirms that a new generation of "digitalized components and systems, that is, Cyber-Physical Systems (CPSs)" are embedded into the industrial environment (Industrial Cyber-Physical Systems (ICPS)), and the technological, economic, and social impacts are being enormous so that the whole process is being labeled as fourth Industrial Revolution (Figure 14.1). The paradigm-shift that this revolution implies is recognized as Industry 4.0 (German: Industrie 4.0) in Germany (see acatech, Industrie 4.0 Working Group 2013; Platform i40 2018; The Federal Ministry of Education and Research (BMBF), Germany 2014).

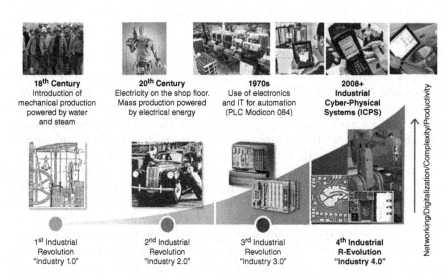

FIGURE 14.1
The industrial environment from the first to the fourth Industrial R-Evolution.

Industry 4.0 is a collective term for technologies and concepts of the whole value chain organization, Combining Cyber-Physical Systems (CPSs), the Internet of Things (IoT), and the Internet of Services (IoS). The major result of this fusion is a modular structured industrial environment where CPSs monitor physical processes, create a digital shadow of the physical world, and perform control and management decisions. Over the IoT, the CPSs structurally communicate and functionally collaborate with each other in real time. Via the IoS, all IoT including humans offer their capabilities as services, both internal and cross-organizational. These services can be subscribed and utilized, in an individual or composed manner, by all IoT participating in the networked value chain (Colombo, Schleuter, and Kircher 2015).

As a matter of fact, manufacturing industry evolved and is now on the edge of this so-called fourth Industrial Revolution. Companies are forced by various initiatives and technological trends, namely Industrial Internet of Things (IIoT), Industry 4.0, and Industrial Cyber-Physical System (ICPS) to move from traditional control and automation approaches toward digitalized, informatized, and networked intelligent manufacturing systems that are architectural reconfigurable, dynamically adaptable (agile) to changing production environment, and flexible to different business opportunities.

The results of the industrial digitalization and as such the amalgamation of CPS, the IoT, and the IoS promise increased industrial competitiveness by enabling shorter product life cycles, increased product variety, shorter time to market as well as shorter time in market, and customized tangible products and services.

14.2 The PERFoRM Approach: Vision and Goal

The vision behind the Production harmonizEd Reconfiguration of Flexible Robots and Machinery (PERFoRM) approach was not to start from scratch but focus on overcoming the last remaining barriers to start exploiting the results and know-how already generated by former innovation projects, as described in Chapter 1. This means, collect the results, methods, and tools of selected previous relevant projects and increase the technology readiness so that those results achieve technology readiness levels 6 and 7, that is, industrialize them. Hereby, the focus does not lie only on technical challenges, but also on non-technological barriers that have hindered the broad industrial implementation of CPS, IIoT, and IoS, so far. A key aspect in this context is the developed migration strategy that holistically supports industries in their journey toward digitalized, harmonized production systems, taking into account technical, economic, and social aspects. In the context described above, the application of the PERFoRM approach supports the introduction of CPS into selected domains of the manufacturing industry and shows with dedicated innovations and Use Cases how to pave the way for manufacturing companies toward Industry 4.0.

More specifically, PERFoRM is dealing with the digitalization, harmonization, integration, prototype implementation, and industrial proof of different technologies to develop true industrial plug-and-produce components and systems suitable for the current and future digitalized industrial manufacturing ecosystem. The approach innovates also in migrating legacy industrial manufacturing automation and control systems to agile plug-and-produce systems in four different manufacturing applications.

One major objective of PERFoRM was to have pilots tested in the real industrial environment. These pilots were specified, developed, implemented, and validated, as described in Chapters 10–13, following four Use Cases that are representative for major key industrial manufacturing sectors: white goods, aerospace, automotive as well as power and gas, respectively. All four pilot Use Cases reached a degree of innovation beyond the "valley of death" confirming the industrial applicability of the solutions. This means, the innovative technology utilized to validate the PERFoRM approach achieved levels 6 and 7 of readiness (according to the international Technology Readiness Level (TRL) scale, NASA (2012), and European Union (2014)).

The industry-like validation pre-test beds provided by Use Cases were used to prove the developed solutions and the readiness levels 4 and 5, before they were implemented in the real industrial environment of the Use Cases. All necessary complementary prototype technologies and associated solutions, such as plug-and-produce device interfaces, plug-and-produce device, and system interoperability checks, were also reaching conceptual integrity as well as industrial readiness at levels 3–5 of readiness.

14.3 Key Challenges on the Way to PERFoRM-Compliant Systems

To truly achieve an extensive and thorough industrial implementation of PERFoRM-compliant solutions, a number of existing barriers and obstacles have to be overcome. Some of these challenges are of technological nature, but many of them have a strong reference to business and economics as well as to mindset and human nature.

The challenges have been described in detail in Chapter 1, but a short recap should be given here to better understand the solution approaches described in Section 14.4.

One of the main barriers for the easy introduction and deployment of the new production systems is the *lack of widely accepted standard plug-and-produce devices*. There is currently a plethora of different realizations of the same basic concepts developed by research projects and local industrial organizations alike. Plug-and-produce is not new, but a lack of common reference architecture and standardized interfaces prevents the formation of larger user communities and negates the expected reduction of integration effort.

Second, *maintaining high levels of quality over small lot sizes* can be considered as one of the most relevant challenges that the manufacturing sector has to face. In fact, any changeover between product variants, even if it appears to be very little, is likely to require a retuning of the production system (re-ramp-up) to ensure that the quality objectives can be consistently achieved. Engineering out any change over variation currently is achieved with sophisticated tools and fixtures that are prohibitively expensive. This restricts the ability of Original Equipment Manufacturers (OEMs) to reduce their lot sizes or adds substantial costs to the components. Soft methods are required that can help to ensure first time right even after production system changes.

Third is the *increasing speed of change*. In a truly global market, the speed of change is increasing while at the same time it is becoming increasingly harder to predict. Forecasting is becoming increasingly difficult, and economic risk of long manufacturing lead times resulting from inflexible production assets, high effort of production changeover, and long distances from the market is becoming increasingly unsustainable.

Agile, plug-and-produce systems require a fundamental change to the engineering approach of automation components, machines, and systems. Advanced optimization and dynamic decision-making models are far more complex than traditional planning and control approaches. Existing skill sets and engineering methods that have been used for many years will either become obsolete or have to be substantially adjusted. This means that there is a lot of *resistance to change* on all levels of the business and technology supply chain that will have to be overcome.

Although a considerable number of countries are highly industrialized, they have many *legacy production equipment and systems* that have been grown and refined over many generations and through several industrial revolutions. While this is clearly strength in view of the fourth Industrial Revolution, it is also becoming an inhibitor for change. A possible obstacle to achieve a real impact from PERFoRM-compliant system implementation could be an incomplete integration of existing production system equipment into the new frame. Two completely split environments (flexible ICT-based and the traditional one) cannot actually work in a fruitful way if they do not share the same basic architecture, do not follow the same production automation concept, do not possess the same interfaces for guaranteeing structural connectivity and behavioral interoperability of legacy and new components, or do not guarantee that the new system components are developed and implemented for being integrated as legacy systems in the future. However, the implementation of new control technologies will have a direct impact on the normal operational status of production while engineers will also face several challenges and obstacles in adopting intelligent automation systems.

Another obstacle could come from the endorsement and missing *acceptance* of the new more flexible production systems *in the industrial domain*. Doubts on the robustness of the approach, the possibility to integrate old and new environments and issues related to human factors as well as a lack of communication could generate skepticism and eventually stop the evolution. Therefore, specific attention needs to be paid to involvement, motivation, and communication to all the stakeholders. Acceptance should not only address technological issues, but also target economic aspects because the missing verification of economic feasibility might be another obstacle for the acceptance.

A *large number of diverse interest groups* are operating on the stage. The automation technology, system integrator, and end user community are notoriously diverse and often lack means to organize themselves. Focus on diverse niche areas and fierce competing make it difficult to agree on wider standards. Strong, influential organizations will have to lead the way that can be pushed down the technology supply chain in one direction and is also open for all actors to participate and strive within their respective areas.

The *techno-economic risk of new paradigm adoption* is too high or at least is not a well-identified reliable approach to forecast real benefits out of large investments. The manufacturing industry is traditionally very careful adopting new technologies especially as a more fundamental change of not only an individual focused area but also the engineering and even a business and a social approach are involved. It will take a consolidated effort of key representative industrial organizations to make the first step to show the benefit for the wider industrial automation and digitalization community. It is the challenge of the first adopter of technology. Everyone is waiting for someone else to make the first step at the moment. Once the benefits have been clearly demonstrated in a real-life industrial and business environment, others will quickly follow. Increasing the Societal Readiness Level (SRL) of the Industrial Digitalization will then be a first major consequence.

14.4 PERFoRM Approach: Key Features and Innovative Solutions

The implementation of the PERFoRM approach not only allowed validating how typical barriers could be overcome from a technical point of view, but also allowed showing economic benefits and better understanding the role of humans in the different digitalized industrial manufacturing domains addressed by the four Use Cases. Table 14.1 summarizes main challenges for a broad application of the PERFoRM approach and shows the key features and innovative solution by which these challenges have been tackled.

14.4.1 Open System Architecture

Instead of relying on only one selected technological standard or providing yet another dedicated technology, the application of the PERFoRM approach did allow developing a set of solutions that is as open as possible to multiple industrial standards and technologies. Principally, the middleware-based approach provides a common ground on which different technologies can interact. Information is exchanged between systems and services by using a common language, PERFoRMML. However, via technology adapters all kind of systems and devices can be connected to the system architecture, independently from their interfaces.

The PERFoRM open system architecture with technology adapters enables a gradual integration of systems and helps to keep "old" legacy systems in operation, that is, legacy systems can be integrated into the system architecture step by step.

14.4.2 Multi-Domain Industrial Application

All PERFoRM solutions were applied in a real industrial environment for each of the four Use Cases. Therefore, technical feasibility could be shown as well as the impact and benefit of the solutions could be made transparent. In addition, prototypical demonstrations that were made in the test beds provided by the Universities and Research Centers did offer excellent opportunities to show all Key Features of the Use Cases, before and after they were running in the real industrial environment.

The Use Cases itself have a large diversity regarding product complexity, lot size, company size, industrial domain, and level of digitalization

TABLE 14.1

PERFoRM Innovative Solutions Approaching Industrial Challenges

Challenges \ Tackled by	Open System Architecture	Multi-Domain Industrial Application	Migration Strategy	Holistic Consideration of Impact	Standardization and Dissemination
Lack of widely accepted standard	●	●	◑	◑	●
Maintaining high levels of quality over small lot sizes	●	○	○	○	◑
Increasing speed of change	●	○	◑	◑	○
Resistance of change	●	◑	●	●	◑
Legacy production equipment and systems	●	●	●	◑	◑
Acceptance in industrial domain	●	●	●	●	●
Large number of diverse interest groups	●	◑	◑	●	●
Techno-economic risk of new paradigm adoption	●	◑	●	●	◑

● Fully addressed ◑ Partially addressed ○ Not addressed

TABLE 14.2

Use Case Characteristics

	Use Case by Siemens	Use Case by I-FEVS	Use Case by Whirlpool	Use Case by GKN
Industry domain	Power and gas	Electric vehicles	White goods	Aerospace
Product type	Industrial compressors	Electric cars	Microwave ovens	Jet engine components
Product complexity	High complexity	Medium complexity	Low complexity	High complexity
Lot size and production type	Small lot size, single-item production	Medium lot sizes, series production	Big lot sizes, mass production	Small lot sizes, series production
Level of automation	Low level of automation	Medium level of automation	High level of automation	Low level of automation

and automation, reflecting a large share of the manufacturing industry (see Table 14.2).

The industrial setups behind the four PERFoRM Use Cases could prove that the developed solutions are not limited to a specific domain but may be applied across various other industrial sectors. The Use Cases did not only demonstrate the technical feasibility of digitalized and harmonized, reconfigurable, and flexible production system concepts, but also the economic benefits of such systems by measuring the quantitative improvements in comparison to the current situation in the real production plants. Moreover, the results of the Use Case applications did allow showing the feasibility and applicability of CPS, IIoT, IoS, and other taylored Industry 4.0-related technologies in industry, as well as a set of considerable benefits for the involved companies. (Note: This has been and is still an important requirement to reduce the resistance to change and to increase the acceptance of the PERFoRM approach among industrial manufacturing companies.)

14.4.3 Migration Strategy

Digitalized and harmonized, reconfigurable and flexible production systems, as envisioned in, for example, the European Factory-of-the-Future program (see EU MANUFUTURE 2013; EFFRA.EU 2018) and the Industry 4.0 initiative (BMWI 2018), are a keystone for economic success of manufacturing industries. But they will only be successful if they can be implemented on the basis of existing architectures and technologies, that is, legacy production management, control, and automation systems.

Hence, an approach has to be adapted which allows the well-managed transformation of existing legacy equipment and systems to become more agile and responsive to changes.

Migration methodologies, tools, and technologies need to be easy to adopt and with a clear cost/benefits rationale. The role and competence of production managers, supervisors, operators, technicians, and other human resources also need to be well-defined and developed in parallel with the technical solutions.

In order to achieve the PERFoRM goals toward the deployment of digitalized, harmonized, reconfigurable, and flexible manufacturing systems, a migration process that allows its implementation in a smoothly manner is needed. A new step-wise migration strategy has been developed and applied to holistically support industries in their journey toward PERFoRM-compliant solution taking into account technical, economic, and social aspects.

This migration process offers to the user guidance, a clear plan to follow, throughout the migration from the legacy into the target system. This makes it possible, for all involved actors of the value chain, to have a better know-how of what to expect next in the migration of the system.

14.4.4 Holistic Consideration of Impact

Holistic consideration means that PERFoRM does not only focus on technological aspects but also explicitly evaluates the impact of digitalized, harmonized, reconfigurable, and flexible industrial manufacturing systems on other aspects such as business, economic, and people, which are often seen as barriers for the successful implementation of innovations like those addressed in this book.

14.4.5 Standardization and Dissemination Activities

As described in Chapters 1 and 2, PERFoRM adopts and develops a large variety of technologies and standards that originate from other initiatives. The developed solution focuses on new forms of reconfigurable plug-and-play capabilities of the production systems as well as on a seamless reference architecture and integration technologies. The described openness of the developed system architecture regarding technical implementation helps not to exclude stakeholders and interest groups.

Moreover, it provided industry proven and reliable standards for the application of the PERFoRM approach.

14.5 New Challenges and Research and Innovation Opportunities

There are opportunities for further application of PERFoRM approaches and technologies within and beyond the manufacturing business.

In the short term, the evaluation of the benefits of applying PERFoRM-compliant solutions will be continued by the innovators in order to also get long-term experience with such systems. This will also impact increasing the acceptance of Industry 4.0 among manufacturing companies.

Mid-term vision is the application of the PERFoRM results beyond the manufacturing business (e.g., process industry). There is still a lack of solutions that support the integration process industry domain such as of casting and molding, forming and coating, and laminating assets.

Finally, the demonstration and multiplication of the PERFoRM integrated solutions across various factories is a long-term opportunity, which will also generate impulses for introducing results into the product and solutions portfolio of industry in other sectors and application domains.

14.6 Final Considerations

For years, several emergent engineering approaches and related Information-Communication-Control Technologies (ICCT), such as Multi-Agent Systems, Service-oriented Architecture, Plug-and-Produce systems, Cloud and Fog technologies, Smart Big Data, Analytics and Metrics, among others, have been researched and prototype developed in a variety of research and innovation activities. The confluence of those results with the latest developments in Digitalized Mechatronics, ICPSs, Systems-of-Systems Engineering, IoT, Industry 4.0, and IoS is opening a new broad spectrum of innovation possibilities for researchers, practitioners, and industrialists.

The PERFoRM approach addressed the development and industrial prototype implementation of a new generation digitalized, agile, and flexible manufacturing system, based among others on plug-and-produce concept and Industry 4.0-compliant technologies, in order to cope with smaller lot sizes and shorter lead time and time to market.

This book reviewed the state-of-the-art methodologies and technologies of flexible, reconfigurable, digitalized, and informatized industrial manufacturing systems outlining the differences between them and the arising benefits within the vision of the PERFoRM approach. It presented new innovations addressing architectural and functional aspects of digitalized industrial manufacturing systems, as well as new engineering methods and tools to support the migration of legacy systems into Industry 4.0-compliant manufacturing infrastructures. It further described a set of test beds and industrial Use Cases in different application domains (electric vehicles, white goods, aerospace, and machine production). They are used to validate the PERFoRM approach and give overviews over current trends and challenges in industrial digitalization in front of the proposed PERFoRM approach, summarizing and out looking new research and innovation opportunities.

The transformation toward flexible, reconfigurable, digitalized, and informatized manufacturing systems following the PERFoRM approach must actively be shaped to fully achieve the promised benefits. This is not only an industrial policy topic but also addresses ecological and social challenges such as resource efficiency, environmental protection, demographic change, urbanization, better work conditions, and a full integration of humans in the new digitalized industrial ecosystem (Bauernhansl, Hompel, and Vogel-Heuser 2014; Colombo et al. 2017; Foehr et al. 2017).

This book aims to contribute to approaching some of these challenges!

References

acatech, Industrie 4.0 Working Group. 2013. "Recommendations for Implementing the strategic initiative INDUSTRIE 4.0." Accessed October 16, 2018. https://www.acatech.de/Publikation/recommendations-for-implementing-the-strategic-initiative-industrie-4-0-final-report-of-the-industrie-4-0-working-group/.

Bauernhansl, T., M. t. Hompel, and B. Vogel-Heuser. 2014. *Industrie 4.0 in Produktion, Automatisierung und Logistik.* Wiesbaden: Springer Fachmedien Wiesbaden.

BMWI, Germany. 2018. "Platform Industrie 4.0." Accessed October 10, 2018. https://www.bmwi.de/Redaktion/EN/Publikationen/plattform-industrie-4-0-digital-transformation.html.

Colombo, Armando W., Dirk Schleuter, and Matthias Kircher. 2015. "An Approach to Qualify Human Resources Supporting the Migration of SMEs into an Industrie4.0-Compliant Company Infrastructure." In *IECON 2015—Yokohama: 41st Annual Conference of the IEEE Industrial Electronics Society: November 9–12, 2015, Pacifico Yokohama, Yokohama, Japan,* edited by Kiyoshi Ohishi and Hideki Hashimoto, pp. 3761–3766. Piscataway, NJ: IEEE.

Colombo, Armando W., Stamatis Karnouskos, Okyay Kaynak, Yang Shi, and Shen Yin. 2017. "Industrial Cyberphysical Systems: A Backbone of the Fourth Industrial Revolution." *IEEE Ind. Electron. Mag.* 11 (1): 6–16. doi:10.1109/MIE.2017.2648857.

EFFRA.EU. 2018. "Factories-of-the-Future Roadmap." Accessed October 10, 2018. https://www.effra.eu/factories-future-roadmap.

EU MANUFUTURE. 2013. Accessed October 10, 2018. http://www.manufuture2013.eu/images/MANUFUTURE2013_Catalogue.pdf.

European Union, Definitions TRL. 2014. "Technology Readiness Levels." Accessed October 15, 2018. https://ec.europa.eu/research/participants/data/ref/h2020/wp/2014_2015/annexes/h2020-wp1415-annex-g-trl_en.pdf.

Foehr, Matthias, Jan Vollmar, Ambra Calà, Paulo Leitão, Stamatis Karnouskos, and Armando W. Colombo. 2017. "Engineering of Next Generation Cyber-Physical Automation System Architectures." In *Multi-Disciplinary Engineering for Cyber-Physical Production Systems: Data Models and Software Solutions for Handling Complex Engineering Projects.* Vol. 8, edited by Stefan Biffl, Arndt Lüder, and Detlef Gerhard, pp. 185–206. Cham: Springer International Publishing.

NASA, TRL. 2012. "Technology Readiness Level." Accessed October 15, 2018. https://www.nasa.gov/directorates/heo/scan/engineering/technology/txt_accordion1.html.

Platform i40, 2018. "Industrie4.0." Accessed October 12, 2018. https://www.plattform-i40.de/I40/Navigation/DE/Home/home.html.

The Federal Ministry of Education and Research (BMBF), Germany. 2014. "Innovationen für die Produktion, Dienstleistung und Arbeit von morgen." Accessed October 19, 2018. http://www.bmbf.de.

Index

Error proofing, 286
Ethernet, 56–57
EU FP6 SOCRADES Consortium
    2009, 58
EU HORIZON 2020 FoF PERFoRM
    Deliverable
  D2.2, 294
  D2.3, 281
  D6.4, 297
  D10.1, 278, 294
EuroNCAP crash tests, 249, 253
European Commission (EC), 46
European Committee for Electrotechnical
    Standardization (CENELEC),
    40, 49
European Committee for
    Standardization
    (CEN), 40, 49
European Telecommunications
    Standards Institute (ETSI),
    40, 44
"ExecutionStatus" OPC-UA tag, 188
Exhaust/ventilation, 290

F

Factory middleware, 279
Factory system architecture, *263*
Failure grouping, 137–138
Failure tickets, 202
Fantini, Paola, 213
Ferreira, Adriano, 213
Fiat, 249
Filippo, Boschi, 1
Finite state automata (FSA)
    formalism, 224
Finite state machine (FSM), 224
5G communication, 56–57
Flexibility, 277
  cluster, 89, **89**
Flexible production, 4, 36, 141, 189, 262,
    307, 308, 311
Floodh, Krister, 275
Floor subframe assembly, 256
Focus Groups on Block chain, 49
Foehr, Matthias, 1, 303
Ford, 248, 249
Forecasting, 4
Forsman, Stefan, 275

Fourth Industrial Revolution,
    304, 305, 308
Frame PWT assembly, 256
Framing geo welding, 258
Frauenfelder, Martin, 99
Front subframe assembly, 256
Functional layer, 39

G

G20, 13
Gartner's Strategic Technology Trends
    for 2016 report, 80
Gateways, 55
  cloud-connectivity, 55
  hybrid infrastructure, 55–56
  legacy integration solutions, 55
Genetic algorithm (GA), 143
Geographic information systems (GIS),
    241
Gepp, Michael, 1, 245, 303
German Commission for Electrical,
    Electronic & Information
    Technologies of DIN and VDE
    (DKE/VDE), 45
German Electrotechnical Society (VDE),
    47
German Institute for Standardization
    (DIN), 45
German Standardization Roadmap
  2018, 48
  2016, 50
GetPMLObjects, 94
Gett, 247
GKN
  architecture, 294, *296*
  micro-flow cell concept, 278, 290
    cell communication, 294–297
    cell overview, 290–291
    cell reconfiguration, 297–299
    industrial context, 278–282
    process modules, 291
    robot system and tooling, 291–292
    safety system, 292–294
    workflow, 282–285
  OPC-UA demonstration and
    validation approach
    results, 188–189
    setup, 185–188